AMBER
The Natural Time Capsule

Andrew Ross

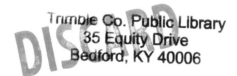
FIREFLY BOOKS

A FIREFLY BOOK

Published by Firefly Books Ltd. 2010

First printing

Publisher Cataloging-in-Publication Data (U.S.)

Ross, Andrew.
    Amber : the natural time capsule / Andrew Ross.
[112] p. : ill., col. photos., maps ;   cm.
Includes bibliographic references and index.
Summary: The book explains how amber is formed, where it is found and how to distinguish genuine amber from fakes. It describes its many uses, both in art and science, and recounts the elusive search for DNA from fossilized insects. Detailed keys and photographs are included to identify species of insects and other amber inclusions.

ISBN-13: 978-1-55407-609-3
ISBN-10: 1-55407-609-9
1. Amber. I. Title.
553.29 dc22   QE391.A5R677 2010

Library and Archives Canada Cataloguing in Publication

Ross, Andrew, 1968-
    Amber : the natural time capsule / Andrew Ross.

Includes bibliographical references and index.
ISBN-13: 978-1-55407-609-3
ISBN-10: 1-55407-609-9

    1. Amber. 2. Amber fossils. I. Title.

QE391.A5R68 2010   553.8'79   C2009-905100-1

Published in the United States by
Firefly Books (U.S.) Inc.
P.O. Box 1338, Ellicott Station
Buffalo, New York 14205

Published in Canada by
Firefly Books Ltd.
66 Leek Crescent
Richmond Hill, Ontario L4B 1H1

Front and back cover images: © NHMPL.

Printed in China

Contents

Preface

I N 1993 I STARTED WORK CURATING the amber collection in the Department of Palaeontology at the Natural History Museum, London. This was in response to interest generated in amber by the film *Jurassic Park*. It was my task to clean, identify and organize the specimens. Books on amber often name the insects and other inclusions that have been preserved but do not explain how to identify them. For this I had to look at descriptions in reference books.

Although there are many characteristics to separate the different groups, most of these features are difficult to see in amber because the insect is not in its natural resting state. Soon I obtained an idea of which features were useful in separating the different groups, although some of these are personal observations that are not listed in entomological books. I also realized which insects are common in amber and those that are rare, which in many cases are completely different to what are rare or common today.

If you buy or find a piece of amber containing an animal, it is likely to be something common such as a fly or ant. This book will help you to identify the creatures you are most likely to see – most are represented by photographs.

Finally, I hope you will learn something new and be able to share in the wonder and appreciation of this remarkable substance.

ANDREW ROSS

ABOVE (fig 2) A piece of Baltic amber with a swarm of fungus gnats preserved in it. (Length of piece 37 mm, or 1 $7/16$ in.)

OPPOSITE (fig 1) Colombian copal with a variety of insects preserved in it. (Maximum width 92 mm, or 3 $5/8$ in.)

CHAPTER 1
What is amber?

OPPOSITE (fig 3) Resin oozing from under the bark of a cedar tree, where a branch has been sawn off. (Width of sawn branch 115 mm , or 4 7/16 in.)

A MBER IS A LIGHT, ORGANIC SUBSTANCE THAT IS usually yellow or orange in colour and often transparent. It is easy to carve and polish, which makes it a popular material for jewellery.

Amber is fossilized resin that once exuded out of the bark of trees (fig 3), although it can also be produced in the heartwood (fig 4). It is not the same as sap, however, which transports nutrients through the heartwood. Resin protects a tree by blocking gaps in its bark. It has antiseptic properties that protect the tree from disease and it is also very sticky and can gum up the jaws of insects that are trying to gnaw or burrow into the bark. Some types of tree can produce a lot of resin, particularly from cracks in the bark or from where branches have broken off. The resin is exuded as blobs or stalactites, which drip and flow down the trunk of the tree. Often, as it exudes, insects become trapped and engulfed in the sticky material. The resin eventually falls to the ground and becomes incorporated into the soil and sediments, and over millions of years it fossilizes into amber. Any insects and other organisms that have been trapped in it are well preserved.

Different types of tree produce different types of resin and in different amounts. Conifers and certain flowering trees produce a lot of resin, particularly in hot weather. The heat also makes the resin less viscous. On a hot day you can watch blobs of resin ooze from cracks in the bark and slowly flow and drip down the trunk. It looks very similar to wax dripping down the side of a candle, although not so fast. Not all tree

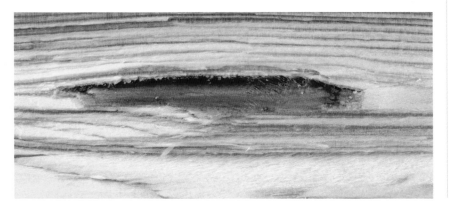

LEFT (fig 4) Piece of pine wood with a resin-filled cavity inside. (Length of resin 52 mm, or 2 in.)

RIGHT (fig 5) Foliage of the legume tree *Hymenaea courbaril* from Dominica, Central America. (Length of black fruit 55 mm, or 2 3/16 in.)

FAR RIGHT (fig 6) Kauri pine (*Agathis australis*) in New Zealand. This species of tree is the source of the resin known as kauri gum.

BELOW (fig 7) Flow of copal from East Africa, produced by *Hymenaea*. (Length 165 mm, or 6 ½ in.)

resins can form amber, as most get broken down and decay. Only two types of tree living today produce stable resins that could, with time, fossilize into amber. They are species of the legume *Hymenaea* (fig 5) in East Africa and South and Central America and the kauri pine (*Agathis australis*) (fig 6) of New Zealand. The resin produced by the kauri pine is called kauri gum, although it is not a true gum.

HOW IS AMBER PRODUCED?

Several factors affect the production of amber from resin, a process known as amberization. Once the resin is exuded it hardens. Resin contains liquids such as oils, acids and alcohols, including the aromatic compounds that produce the distinctive resinous smell – two examples of highly aromatic resins are frankincense and myrrh. Scientists call these liquids volatiles and they dissipate and evaporate from the resin. The resin then undergoes a process known as polymerization, whereby the organic molecules join to form much larger ones called polymers. Hardened resin is known as copal (figs 7, 8). Copal becomes incorporated into soil and sediments, where it remains long after the tree dies. It continues to polymerize and lose volatiles until the resultant amber is completely polymerized, has no volatiles and is inert.

Many scientists thought that time was important in the fossilization of resin to produce amber, and the amberization process was estimated as taking between 2 and 10 million years. However, it now appears that many more factors are involved. Most amber in deposits around the world was not formed where it is found – the copal or amber has been eroded from the soil, transported by rivers and deposited elsewhere. For instance, amber from Borneo is 12 million years old and comes from sand and clay sediments that were deposited in a deep ocean. The fossilized resin from Borneo that comes from beds of sandstone is completely inert and undoubtedly amber. However, resin that comes from beds of clay still contains volatile components, which means that it is still copal. So, the type of sediment in which the resin is deposited is much more important than time for amber formation. But what is not so clear is the effect of water and sediment chemistry on the resin.

CHEMICAL PROPERTIES

Amber is described as an amorphous, polymeric glass (see Glossary). The polymers in ambers are cyclic hydrocarbons called terpenes. A familiar terpene is turpentine, a volatile liquid obtained from tree resin, which is often used as a solvent. Amber generally consists of around 79% carbon, 10% hydrogen and 11% oxygen, with a trace of sulphur. Amber can be partially dissolved in organic solvents, but it is not affected by most alcohols.

PHYSICAL PROPERTIES

Amber has a hardness of 2–3 on the Mohs scale, which means it is not very hard (about as hard as a fingernail) and can be easily scratched. It has a specific gravity (s.g. or relative density) of 1.04–1.10, which is only slightly heavier than water (s.g. of 1.00). Some pieces of amber contain air bubbles, which lower the specific gravity and allow the pieces to float – amber is readily transported by rivers and tides. It has a melting point range of 200–300°C (392–572°F), but it tends to turn black and burn, rather than melt. It may fluoresce blue under ultraviolet light and will produce static electricity if rubbed. Amber is warm to the touch and, if broken, can produce a conchoidal fracture, which looks like the surface of a shell.

VISUAL CHARACTERISTICS

Amber is very popular for making jewellery and there is considerable interest in the insects and other objects, or inclusions, trapped in it, particularly large and rare ones. Because of this, amber is valuable and often faked. However, fakes usually have tell-tale features that enable them to be identified visually without resorting to damaging testing.

A lot of modern amber jewellery contains circular, radial cracks called sun spangles (fig 9), which are made by heating and cooling the amber. Green amber jewellery with sun spangles has recently become common, but this colour is not natural. The amber has either been subjected to heat and pressure with a gas that has altered its colour, or the back of the amber is burnt or painted black and when placed in a silver mount it looks green. The sun spangles used to be a good indicator of genuine amber, but in recent years forgers have found a way of reproducing the cracks in plastic.

Insect inclusions that are only a couple of millimetres long are likely to be genuine, as small insects in amber are common. Natural features in Baltic amber, that are not faked, are the presence of the hairs from the flowers of oak trees and black cracks filled with tiny pyrite crystals. Larger genuine insects in Baltic amber are often partly

ABOVE (fig 8) Stalactite of New Zealand copal (kauri gum). (Length 180 mm, or 7 1/16 in.)

BELOW (fig 9) Pendant containing circular cracks known as sun spangles. These cracks are common in jewellery and are artificially produced by heating the amber. (Length excluding rings 60 mm, or 2 3/8 in.)

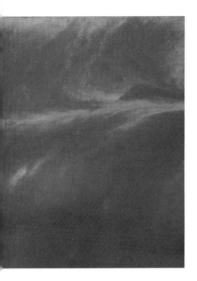

ABOVE (fig 10) Pressed amber, also known as ambroid. This substance is formed by fusing small pieces of amber together. (Length 50 mm, or 1 ¹⁵/₁₆ in.)

covered in a white coating, which is a product of decay. These features indicate that the specimen is genuine, although their absence does not indicate that it is fake, as many pieces of amber are perfectly clear and without any inclusions.

Carved or drilled forgeries, where a piece of genuine amber is hollowed out and an inclusion is inserted, can only be identified visually. Any tests carried out on the surrounding amber would indicate that it is genuine. The inclusions used tend to be large (more than 10 mm, or about ⅜ in, long). Knowledge of modern animal groups and those found in amber is useful in identifying fakes such as these, as only a few insects recorded from amber belong to living species – the rest are extinct. A faked inclusion will belong to a living species and usually only a specialist can identify it. Close examination under a microscope should reveal planes running through the amber where it has been cut and hollowed out. However, there are also natural cracks and flow planes in amber that look similar.

Pressed amber, also known as ambroid, is commonly found in Victorian jewellery and as the stems of tobacco pipes. It is formed by fusing small pieces of amber together under high temperature and pressure. It can be clear or cloudy, or a mixture of clear and cloudy swirls (fig 10), and it can be a variety of colours. Yellow and orange pressed amber is very difficult to distinguish from unadulterated amber. It cannot really be called fake amber because it is made out of genuine amber.

THE PILTDOWN FLY

A carved fake discovered in the collection of the Natural History Museum, London, is now famous and dubbed 'The Piltdown Fly' (fig 11). The specimen consists of a large housefly in a piece of Baltic amber and was first mentioned by the German entomologist Hermann Loew in 1850. Much later, in 1966, the famous German entomologist Willi Hennig examined it and identified it as belonging to the living species *Fannia scalaris*. This was important because *F. scalaris* is a very advanced fly and suggested that Baltic amber did not accurately reflect the fauna at that time, because other living species should also be preserved in it. Then, in 1993, the specimen was re-examined using a binocular microscope and Anglepoise lamp. The lamp put out some warmth, which caused a crack to suddenly appear around the fly. This was initially worrying but also suggested that there was something amiss. Close examination from the side revealed two planes, one running flat through the amber and the other around the fly. Clearly, someone had cut a piece of amber in half, hollowed out one side, inserted a fly and glued it back together, The fly has bright red eyes and an abdomen broken in an unnatural way, which also implicated it. Fakes like this are rare, as copal and plastic are more commonly used for inserting inclusions.

ABOVE (fig 11) 'The Piltdown Fly'. This specimen is a carved fake in which a modern housefly has been inserted into a piece of Baltic amber. You can see lines around the fly where the amber has been cut in half and one side hollowed out. (Length of fly 7 mm, or ¼ in.)

FAKE AMBER

Several substances are used to imitate amber, but there are a number of tests that can distinguish between them. The following substances are commonly encountered as fake amber:

- copal
- glass
- phenolic resin
- celluloid
- casein
- modern plastics.

Phenolic resin, celluloid and casein are often seen in Victorian, Edwardian or Georgian necklaces (fig 12), whereas other plastics are used for more modern necklaces. Only copal and modern plastics are used to embed fake inclusions.

COPAL

Copal is often sold as amber because it naturally contains insects and other inclusions. If it is exposed to light and air it degrades to produce a network of small polygonal cracks (crazing) on the surface (figs 13, 14). This process also happens to amber but takes much longer and the amber turns dark orange, whereas copal remains yellow. Copal with natural inclusions usually comes from East Africa, Colombia and the Dominican Republic. Most pieces are light yellow in colour and transparent. Some pieces from East Africa and Colombia can also be colourless or orange (see fig 1), and some pieces from Colombia and the Dominican Republic also have a brownish tinge.

ABOVE (fig 12) Composite necklace containing fake amber beads. The large yellow-orange bead is celluloid, the oval red beads are phenolic resin, the oval cloudy yellow beads are casein, the clear orange beads and small faceted beads are glass. (Length of largest red beads 27 mm, or 1 1/16 in.)

FAR LEFT (fig 13) East African copal with characteristic crazed surface caused by oxidation. The inclusion is a beetle. (Length of specimen 70 mm, or 2 3/4 in.)

LEFT (fig 14) New Zealand copal (kauri gum) with crazed surface. (Length 145 mm, or 5 11/16 in.)

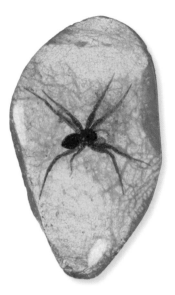

The pieces can contain many different types of insect (fig 16). If some of the insects are only a couple of millimetres long, then the pieces are likely to be genuine. Again, knowledge of the differences between modern insects and those found in amber is useful, as most insects in copal belong to living species. One good indicator is the proportion of flies with long or short antennae. Flies with long antennae are very common in amber whereas those with short antennae are rare. The opposite is true for the modern fauna – most flies trapped in copal have short antennae. There are several specimens in the Natural History Museum, London, that were labelled 'amber' but are actually East African copal. Many were originally shown as amber in the *Quarterly Journal of Science* in 1868. Another specimen labelled as Baltic amber was identified as belonging to the living cockroach species *Euthyrrapha pacifica* in 1911, but this again is in East African copal and not amber.

ABOVE (fig 15) Spider that has been inserted into melted kauri gum. Its large size and outstretched legs are tell-tale signs that it is a fake. Genuine spiders in amber usually have their legs tightly curled up under their bodies. (Length of piece 80 mm, or 3 1/8 in.)

RIGHT (fig 16) East African copal with insects. The three large flies belong to a family that has not been found in amber. (Length of specimen 53 mm, or 2 1/16 in.)

FAR RIGHT (fig 17) Iridescent beetle that has been inserted into melted copal. This species only lives in Asia whereas the copal probably came from East Africa. (Length of piece 19 mm, or 3/4 in.)

Faked inclusions are also found in copal. Copal has a much lower melting point than amber (less than 150˚C or 302˚F) and melts rather than burns, allowing organisms to be inserted. Most copal fakes are large (more than 20 mm, or 13/16 in, in length) consisting of lizards, large crickets, spiders and butterflies. Often, the fakes contain only one large inclusion, which is in the centre of the piece with its legs neatly arranged (figs 15, 17). Kauri gum was commonly used as a medium for faked inclusions in the early twentieth century. It is usually transparent and either yellow, orange or red with a crazed surface. Natural inclusions in kauri gum are very rare as few insects live on the kauri pine.

GLASS

Glass is easy to distinguish from amber because it is cold to the touch, dense and does not scratch with steel. See fig 12 for examples of fakes using glass and the next three substances.

PHENOLIC RESIN

Phenolic resin, which is also known as bakelite, is the most common material to be encountered in fake amber necklaces. The necklaces are usually made up of very large oval beads that are smooth or faceted. The beads get progressively larger the further they are away from the catch. There are two colours that are commonly encountered. The most common is a dark red, with every bead virtually the same colour. The beads can be transparent or cloudy. Phenolic resin beads can also be cloudy yellow with swirls, which makes them harder to distinguish from genuine or pressed amber and requires physical or chemical analysis. Phenolic resin is slightly denser than amber.

CELLULOID

Celluloid (cellulose nitrate) is usually yellow and cloudy. It is difficult to distinguish from amber except that it is slightly denser and much more inflammable.

CASEIN

Casein is a plastic made from milk. The beads are cloudy and a dirty yellow colour, and casein is also slightly heavier than amber.

MODERN PLASTICS

Modern plastics such as polyester and polystyrene are commonly used today to fake amber and produce faked inclusions. These are harder to detect because some forgers can produce perfectly transparent pieces that are amber coloured. They are commonly used in Mexico and the Dominican Republic. As with copal, the faked inclusions are usually obvious, because the inclusion is large (usually more than 10 mm, or ⅜ in, long), neatly centred in the piece and belongs to a living species (fig 18). Some forgers also add dirt to try to make a piece look more authentic.

In Mexico, green amber does occur but is extremely rare, yet necklaces and pieces made of green plastic are commonly sold as amber. In the Dominican Republic genuine lizards and scorpions in amber do occur but they are extremely rare and very valuable. However, it is very easy to visit there and buy a fake lizard or scorpion in plastic and pay a lot of money for it. The local lizard population is probably under threat from amber forgers. Plastic has also turned up in carvings that are supposed to be Chinese amber. With the advent of the internet there is much fake amber being sold (as well as genuine pieces), so you have to be careful what you buy.

BELOW (fig 18) Lizard that has been inserted into plastic. Fakes like this are very common because genuine lizards in amber are extremely rare and valuable. (Length 75 mm, or 2 ¹⁵⁄₁₆ in.)

TESTS FOR AMBER

There are several tests that can easily be done to try to determine whether a specimen or necklace is genuine amber. These tests are potentially destructive and should only be used as a last resort if it is not obvious visually whether the piece is genuine or fake. For a summary of these tests see Table 1 below.

THE ALCOHOL TEST

An effective test to distinguish between amber and copal is the alcohol test. This involves putting a drop or two of alcohol, either isopropanol or ethanol, on a polished surface of the specimen and letting it evaporate. The volatile components that are present in copal react with the alcohol, which removes the polish and makes the surface sticky. If you press a finger against the spot a fingermark is often left on the surface of the copal (this can be polished away by rubbing it hard on a piece of cloth). There is no reaction with amber or the other fake materials and the polished surface is usually unaffected.

THE SCRATCH TEST

A good test to distinguish between amber and glass is to try to scratch it with a pin. If it does not scratch then it is glass (hardness of 5–6 on the Mohs scale). If it is amber or the other fake materials it will scratch, so choose part of the specimen where it is not noticeable and try to make the scratch as small as possible. Close examination of the specimen beforehand may indicate that it is not glass, without the need to use this test. If the surface is crazed then it is not glass. Glass can be scratched by diamond (hardness of 10 on the Mohs scale), so if the piece is already scratched this does not necessarily mean it is not glass.

THE HOT WIRE TEST

The hot wire test can be used to distinguish between amber and fake materials. The test involves heating up a wire or needle red hot, allowing it to cool slightly and gently pressing the tip against the surface, which produces a puff of smoke. Sniffing the smoke can indicate what it is. Amber has a slightly acrid resinous smell and copal has a sweet resinous smell, whereas many artificial materials have an acrid plastic-like smell. The wire makes a mark on the surface, so choose part of the piece where it will not show. Take care when using this test because celluloid is highly inflammable and other plastics give off potentially harmful vapours.

THE SALTWATER TEST

Another test is to use a saturated salt solution to see if a specimen will float or sink.
- Fill a glass with 284 ml (half a pint) of water.
- Add seven heaped teaspoons of table salt then stir. The water will initially turn cloudy, then gradually become clear as the salt dissolves.
- Leave it for several minutes, giving it a quick stir every 30 seconds.

BELOW Table 1. The results of four tests on amber and substances used to fake it

A Does alcohol make it sticky?
B Can it be scratched?
C Does it float in a saturated salt solution?
D Does a hot wire produce a resinous smell?

	A	B	C	D
Amber	N	Y	Y	Y
Copal	Y	Y	Y	Y
Glass	N	N	N	N
Phenolic resin	N	Y	N	N
Celluloid	N	Y	N	N
Casein	N	Y	N	N
Other plastics	N	Y	N	N
Polystyrene	N	Y	Y	N

(Y = yes; N = no)

- To test a necklace, remove one of the beads from the string and drop the bead into the glass. The salt solution has a higher specific gravity (1.1) than amber and copal so these materials will float, whereas glass, phenolic resin, celluloid, casein and some other plastics are denser than water and will sink. Polystyrene, however, has the same specific gravity as amber and will float.
- Always thoroughly wash the piece in water afterwards.

This test is of no use for amber pendants with a metal clasp as the weight of the clasp will make the pendant sink. Do not use this test if there are any cracks in the specimen as there is a danger that the saltwater will penetrate the crack and, when it dries out, could produce salt crystals that enlarge the crack.

ANALYTICAL TESTS

There are several scientific tests used to identify amber and the two most often employed are infrared spectroscopy and mass spectrometry. Both produce a graph with many peaks reflecting the chemical composition of the sample. From studying the graph it is easy to find out whether the sample analyzed is amber or one of the fake materials. Infrared spectroscopy can indicate whether the amber is Baltic amber or not, as Baltic amber has a characteristic plateau on the side of one of the peaks, known as the 'Baltic shoulder' (fig 19). Mass spectrometry is able to distinguish between the different types of amber, but this process is more expensive and not widely available.

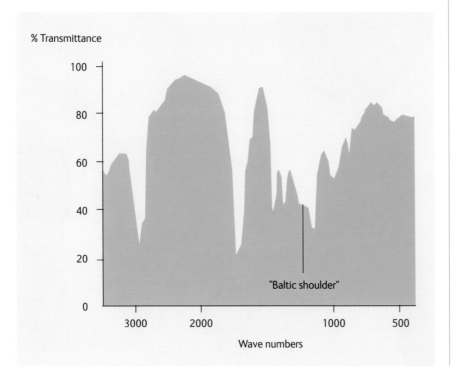

LEFT (fig 19) Infrared spectrograph of Baltic amber, showing the characteristic Baltic shoulder.

THE USES OF AMBER

The most common use for amber is in jewellery. It is popular because it is light, warm to the touch and beautiful to look at. Most amber jewellery consists of pendants, necklaces, earrings, brooches and rings (figs 20, 21, 22).

Because amber is fairly soft, it is easy to carve. Since prehistoric times it has been carved into a variety of ornaments, mainly of people or animals (figs 23, 24, 25), some of which had a significant meaning. It has also been used for everyday items such as bowls, cups, bottles, snuffboxes, knife handles, candlestick holders and quill pen holders. It has been used for recreational pursuits such as games boards, and even for chessmen and dice (fig 27). Pieces have been skilfully inlaid in the surface of various items of furniture including chests, caskets and cabinets – even a chair has been made out of it. Amber has been used a lot for religious items. The main use in medieval times was for rosaries; later it was used in altars and crucifixes. Amber has also been used for pipe stems (fig 26) and cigarette holders, although ambroid is more commonly used for these because it is slightly tougher. More information on the many uses of amber can be found in the books listed on p. 112.

ABOVE (fig 20) A beautiful Baltic amber pendant surrounded by diamonds, containing a spider and a cricket. (Length excluding ring 43 mm, or 1 ¹¹⁄₁₆ in.)

RIGHT A (fig 21) A Baltic amber ring containing a long-legged fly. (Length of ring 19 mm, or ¾ in.)

FAR RIGHT (fig 22) Necklace and earrings made out of Baltic amber. The tiny beads are glass. (Length of the largest piece of amber in the necklace 41 mm, or 1 ⅝ in.)

BELOW (fig 23) Plastic replica of a Stone Age horse carved out of a piece of Baltic amber from Poland. (Length 115 mm, or 4 ⁷⁄₁₆ in.)

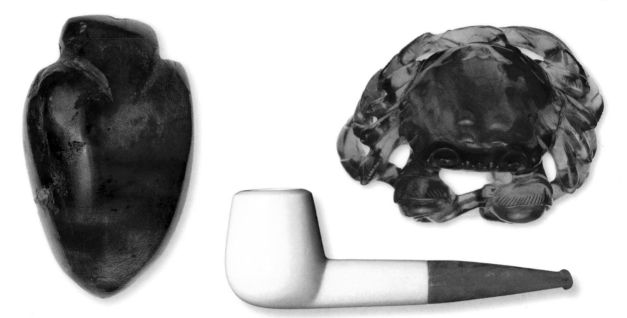

THE AMBER ROOM

The most remarkable use of amber was when it was used to decorate a whole room. The famous amber room was commissioned by King Frederick I of Prussia in 1701. It was made from large panels inlaid with carefully shaped pieces of amber to form a mosaic. In 1717, the room was given to Czar Peter the Great of Russia and placed in the Old Winter Palace in St Petersburg. In 1755, it was moved to the Ekaterininsky Palace in Tsarskoye Selo. During the Second World War, in 1941, the Germans invaded, dismantled the room and took it to Koenigsberg Castle. Since 1945 its exact whereabouts are unknown. There are a number of theories: it was either destroyed by bombing, was hidden in a mine or lake somewhere, or is on a sunken boat at the bottom of the Baltic Sea. The Russians have faithfully re-created the room, based on old photographs, in the Catherine Palace in St Petersburg and it opened in 2003.

AMBER PRODUCTS

Apart from decorative objects, amber has had other uses. The poorer grades of amber in the past were distilled to make colophony, amber acid (succinic acid) and amber oil. Colophony was dissolved in turpentine and linseed oil to make varnish. Amber varnish is hard, slow drying and dark, and was used for stringed instruments, horsedrawn carriages, ships' decks and in early photography. Copal and other natural tree resins were also used to make varnish in the past, whereas nearly all varnishes that are produced today are synthetic. Amber acid was used to produce soap, bath salts, pharmaceutical products, dyes and photographic chemicals, while amber oil was used in wood preservatives and insecticides. Both the acid and the oil were also used in the iron industry.

ABOVE LEFT (fig 24) Ancient artefact of Sicilian amber, showing a crude carving that would have been used as a talisman. (Length 65 mm, or 2 9/16 in.)

ABOVE CENTRE (fig 26) Meerschaum pipe with Baltic amber stem. (Length 103 mm, or 4 in.)

ABOVE RIGHT (fig 25) Crab carved out of Chinese amber. (Length 51 mm, or 2 in.)

BELOW (fig 27) Pair of Baltic amber dice. (Width 12.5 mm, or ½ in.)

CHAPTER 2

Where is amber found?

MBER OCCURS IN MANY PARTS OF THE WORLD, but most of the deposits are small and localized. Most of it is not of a high enough quality to make jewellery, but large deposits of jewellery quality amber are exploited commercially and most comes from the Baltic region and the Dominican Republic. All deposits are of interest to scientists who study their chemistry, while other scientists are interested in the deposits of amber that contain insects and other inclusions. The main insect-bearing ambers come from the Baltic region, the Dominican Republic, Burma (Myanmar), Mexico, Lebanon, Siberia, Canada, New Jersey, Spain, France and Sicily. Insects have also been recorded in amber from Bitterfeld (Germany), China, Japan, Romania, Sakhalin Island (Russia), Claiborne (Arkansas, USA), Alaska, Borneo and the Isle of Wight (UK). Recently, a few insects and other inclusions were discovered in amber from Australia and India. Figure 29 shows where these ambers occur and fig 30 indicates their ages.

OPPOSITE (fig 28) Piece of highly polished oxidized red Baltic amber. (Length of spider 10mm or ⅜ in.)

BELOW (fig 29) Map of the world showing the main amber and copal localities.

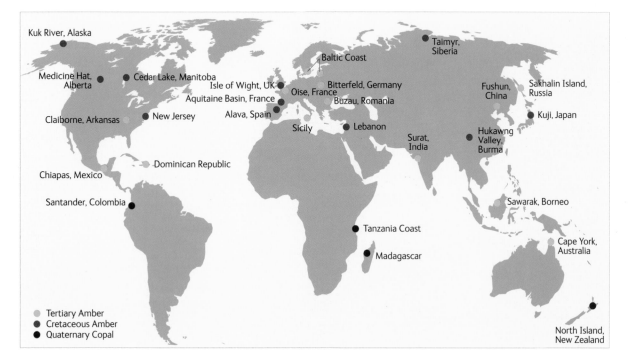

Kuk River, Alaska

Taimyr, Siberia

Baltic Coast

Medicine Hat, Alberta

Cedar Lake, Manitoba

Isle of Wight, UK

Bitterfeld, Germany

Aquitaine Basin, France

Oise, France

Buzau, Romania

Fushun, China

Sakhalin Island, Russia

Claiborne, Arkansas

New Jersey

Alava, Spain

Sicily

Lebanon

Kuji, Japan

Surat, India

Hukawng Valley, Burma

Chiapas, Mexico

Dominican Republic

Santander, Colombia

Tanzania Coast

Sawarak, Borneo

Madagascar

Cape York, Australia

● Tertiary Amber
● Cretaceous Amber
● Quaternary Copal

North Island, New Zealand

RIGHT (fig 30) Geological time column showing some moments in evolution and the ages of some insect-bearing ambers. (Pliocene, 2-5 million years ago and Quaternary, 0-2 million years ago divisions are not labelled.)

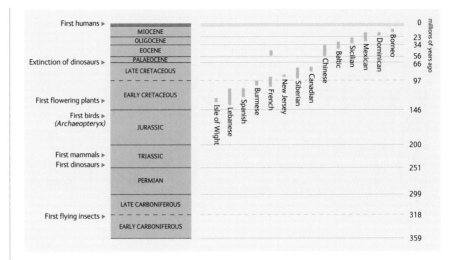

RIGHT (fig 30) Geological time column showing some moments in evolution and the ages of some insect-bearing ambers. (Pliocene, 2-5 million years ago and Quaternary, 0-2 million years ago divisions are not labelled.)

DATING AMBER

Amber can only be dated based on fossils in the associated sediments, as there is no way of knowing how long the amber took to become deposited. If the amber is re-worked (eroded from one deposit and re-deposited elsewhere) then the amber could be much older than the sediments suggest.

The oldest recorded amber comes from the Late Carboniferous Coal Measures. Amber is also recorded from the Permian, Triassic and Jurassic periods. Plant and fungal inclusions have been reported from these early ambers but, as yet, no insects have been found. The oldest known insect-bearing ambers come from the Early Cretaceous. The Cretaceous period was a very important time in the history of life on this planet — the flowering plants radiated and displaced the old conifers, cycads and ferns to become the dominant plants. At the end of the Cretaceous an extinction event occurred that wiped out the dinosaurs.

SOURCES OF AMBER

Most amber is not of commercial interest so is unlikely to be encountered. Of the commercial ambers, Baltic and Dominican are the best known and most readily available. A lot of research has been undertaken recently on Burmese amber (see p.25). Some others do turn up from time to time and these are briefly described below.

BALTIC AMBER

The most common amber in western Europe comes from the countries surrounding the Baltic Sea, namely Poland, Russia, Germany, Denmark and Lithuania (fig 31). It comes from a deposit known as the Blue Earth, which lies below the water table and extends out into the Baltic Sea. Storms can rip out amber from this bed, which dates from the Late Eocene–Early Oligocene, and wash it up on to the shores. Lumps of this amber are also occasionally washed up on the east coast of Britain in Kent,

Suffolk, Norfolk, Yorkshire and even Fife after a storm (figs 32, 33). The largest Baltic amber deposits are on and around the Samland Peninsula, in an area that forms a small part of Russia between Poland and Lithuania. Before the Second World War this area was known as East Prussia and formed part of Germany. The main city there is Kaliningrad, which was previously called Koenigsberg.

When fresh, Baltic amber is usually lemon-yellow or orange and can be cloudy or clear (figs 34, 35, 36). The clear amber often contains insects, whereas the cloudy amber only rarely contains insects. However, because it is opaque, it isn't possible to see inside it. The cloudiness is caused by millions of microscopic air bubbles. A high proportion of bubbles can make a piece white and this is known as bone amber. With time – perhaps at least 50 years – the clear amber oxidizes to dark orange or red due to exposure to air or seawater (figs 28, 37). After a much longer time it develops a cracked crust such as is seen in some archaeological specimens.

Most Baltic amber contains a lot of succinic acid (3–8%) and has been given the name succinite. This acid produces the characteristic 'Baltic shoulder' on the infrared spectrograph (see fig 19, p.15). Other ambers that do not contain succinic acid are known as retinites.

ABOVE (fig 31) Map of the area around the Baltic Sea.

LEFT (fig 32) Large piece of red Baltic amber washed up on Cromer beach, Norfolk, UK. (Length 205 mm, or 8 1/16 in, weight 793 g, or 28 oz.)

BELOW (fig 33) Large polished piece of Baltic amber dredged off the coast at Yarmouth, Norfolk, UK. (Length 170 mm, or 6 11/16 in, weight 1,048 g, or 37 oz.)

There are a few ways of visually identifying clear Baltic amber. It commonly contains tiny hairs (1–2 mm, or about ¹⁄₃₂–³⁄₃₂ in, long) that probably came from the male flowers of oak trees (fig 38), although the resin itself came from conifers. The presence of these hairs indicates that most of the clear resin was produced during the spring and summer, while the oaks were flowering. Often, Baltic amber also contains cracks, some of which are black and filled with lots of tiny pyrite crystals (fig 39). Some

RIGHT (fig 34) Polished piece of cloudy Baltic amber. (Length 55 mm, or 2 ³⁄₁₆ in.)

CENTRE LEFT (fig 35) Unpolished piece of Baltic amber from the Blue Earth, Kaliningrad, Russia. (Length 70 mm, or 2 ¾ in.)

CENTRE RIGHT (fig 36) Polished piece of fresh Baltic amber. (Length 70 mm, or 2 ¾ in.)

BELOW (fig 37) Piece of oxidized orange Baltic amber. The curved, smooth ridges on the surface are where the piece has broken. This is known as a conchoidal fracture. Part of the natural unpolished crust can be seen, represented by small polygonal cracks. (Length 85 mm, or 3 ⅜ in.)

ABOVE LEFT (fig 38) Clusters of tiny hairs, probably from the male flowers of oak trees (Length of the long hair in the centre 1.5 mm, or 1/16 in.)

BELOW LEFT (fig 39) Caddis-fly covered with a white coating formed as a product of decay. On either side of the insect is a crack filled with crystals of iron pyrites (pyrite). (Length 9 mm, or about 3/8 in.)

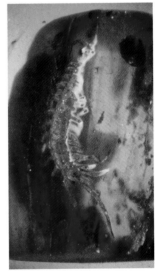

ABOVE (fig 40) Centipede in Baltic amber. This centipede has been half polished away and is filled with pyrite crystals. (Length of piece 19 mm, or 3/4 in.)

of the inclusions are also preserved in pyrite (fig 40). The oak hairs and cracks filled with pyrite crystals are rarely found in other ambers. The insect inclusions are often preserved with a white coating, which is formed from liquids that escaped from the decaying insect's body and then entered the surrounding amber, turning it cloudy (see fig 39). You can only find evidence for this process in Baltic amber.

There has been a lengthy debate as to what type of tree produced Baltic amber. There is no doubt that it was produced by a conifer, but what kind? The extinct amber-producing tree was named *Pinites succinifer* in 1836 by H R Goeppert based on bark inclusions in the amber. It was later placed in the living genus *Pinus* (Family

Pinaceae). More recent chemical analysis indicated that Baltic amber is more similar to the resin produced by monkey-puzzle trees (Family Araucariaceae, which includes the kauri pine), particularly due to the presence of succinic acid. However, some trees within the pine family can also produce succinic acid, and araucarian inclusions are much rarer than pine inclusions. The Baltic amber producing tree had features of both pines and monkey-puzzle trees. The plant and insect inclusions preserved in Baltic amber indicate that it was produced in a subtropical forest with a mixture of different trees.

Baltic amber has a long history. Amber artefacts have been recovered from many archaeological sites. The oldest British finds come from Gough's cave in Cheddar (around 10,000 BC) and Star Carr in Yorkshire (around 8,000 BC) (fig 41). This amber probably originated from the English coast. One of the most important archaeological finds is a Bronze Age amber cup that was discovered in a tumulus grave in Hove, Sussex, UK (fig 42). Other finds occur in countries that are a long way from the source of Baltic amber, indicating that trade in this substance started very early on. Certainly by 1,600 BC, a trade route had opened up to Greece.

In medieval times amber was collected by local peasants from the shore around the Samland Peninsula or netted from the sea itself. It wasn't until the 1850s that systematic mining began in the area with the formation of the Stantien & Becker company. They started by dredging the sea, then went on to open-cast mining followed by underground mining on the Samland Peninsula. The company established a museum to house a large collection of amber, which was maintained by Professor Richard Klebs. In 1899, the company was purchased by the Prussian state and most of Klebs's collection passed to Koenigsberg University. In 1892, two hundred of Klebs's specimens were purchased by the Natural History Museum, London. Commercial open-cast mining on the Samland Peninsula continues to this day. More information on the history of amber and amber mining can be found in the books listed on p.112.

ABOVE (fig 41) Stone Age amber artefact from Yorkshire, UK. This piece is about 10,000 years old (8,000 BC). (Length 23 mm, or ⅞ in.)

RIGHT (fig 42) Bronze Age amber cup from Hove, East Sussex, UK. (Diameter 90 mm, or 3 9/16 in.)

BURMESE AMBER

Burmese amber (called burmite) from Burma (Myanmar) is generally orange or red in colour. Pieces often contain insects and plant debris but they are usually not as well preserved as in Baltic amber. A piece at the Natural History Museum, London, is the largest piece of Burmese amber in the world (fig 43), and was purchased in China in 1860. Another piece (fig 44) is one slice of seven from a large piece, which contained over 450 arthropod inclusions – the richest in the world. Burmese amber was traded with China from at least the first century AD until the British took control in 1885. It was mined from pits in the Hukawng Valley in northern Burma and over 82 tonnes were extracted from 1898 to 1940. The only collection of insects in Burmese amber to be made at this time was one acquired by the Natural History Museum, London, in the 1920s. The amber was thought to be Eocene in age, based on accompanying marine microfossils; however, a detailed examination of the collection in the 1990s showed that the insects were very different to those in Baltic amber and some belonged to extinct families that are not known more recently than the Late Cretaceous (fig 45). It was realized that some pieces had been re-worked and that this amber was probably Cretaceous in age. Burma became independent in 1947 and changed its name to Myanmar in 1989, but has only recently started exporting Burmese amber. This has generated a lot of new research on this fascinating amber. Large collections have recently been built up by the American Museum of Natural History, New York, and private individuals, and these have yielded much new information. The bed that this amber came from has recently been dated as late Early Cretaceous in age.

Burmese amber formed at a time of a major change in life on Earth. The flowering plants were radiating rapidly, which had a major impact on the insect fauna. This is reflected in the amber as it contains a mixture of extinct and living families, and over 150 have now been recorded. The extinct families consist of some that are ancient and had previously only been found as older fossils in rock (fig 46), and others that are unique to Burmese amber (fig 47). For some of the living families these inclusions are the oldest known fossil representatives in the world and have also revealed other

LEFT (fig 43) The largest piece of Burmese amber in the world. (Length 500 mm, or 19 ¹¹⁄₁₆ in.)

RIGHT (fig 44) Large slice of Burmese amber containing plant debris and nearly 200 insect inclusions. (Length of piece 100 mm, or 3 $^{15}/_{16}$ in.)

RIGHT (fig 45) Parasitic wasp (Hymenoptera: Serphitidae) in Burmese amber. This family is only known from Cretaceous ambers. (Length 3.5 mm, or $^1/_8$ in.)

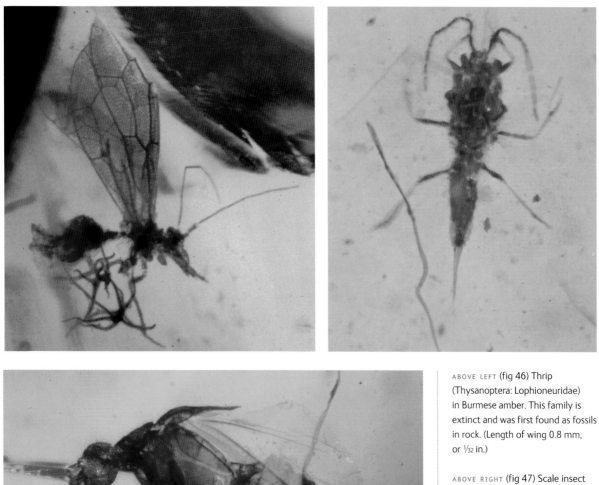

ABOVE LEFT (fig 46) Thrip (Thysanoptera: Lophioneuridae) in Burmese amber. This family is extinct and was first found as fossils in rock. (Length of wing 0.8 mm, or ¹⁄₃₂ in.)

ABOVE RIGHT (fig 47) Scale insect (Hemiptera: Burmacoccidae). This family is only known from Burmese amber. (Length 1.3 mm, or ¹⁄₁₆ in.)

LEFT (fig 48) Wedge-shaped beetle (Coleoptera: Ripiphoridae) in Burmese amber, the oldest known fossil record of this living family. (Length 2.9 mm, or ⅛ in.)

interesting information. Figure 48 is a wedge-shaped beetle and the oldest known fossil of its family, which are specialist parasites of cockroaches. Figure 49 is one of the oldest ants in the world yet it has specialized jaws adapted for lifting objects rather than grasping them, thus indicating that ants probably evolved earlier on (there are also ants known of the same age from French amber). Figure 50 is a mayfly and is the only known fossil of its particular family, which as adults only live for a few hours. Mayflies can have two adult stages – subimago and imago – and this is an imago female, but the living females of this family only get to the subimago stage today.

BELOW (fig 49) Head of an ant (Hymenoptera: Formicidae) in Burmese amber, one of the oldest ants known. (Length of antenna 1.5 mm, or ¹⁄₁₆ in.)

Several species of digger wasps, which are pollinators of flowers, have been described from Burmese amber. There are nine female individuals of the same species (fig 51) trapped in one piece in the collections of the Natural History Museum, London, thus demonstrating early social behaviour. A bee has also been described from Burmese

amber, and is the oldest in the world. Biting insects, including mosquitoes, midges and sandflies, are known in Burmese amber. As this amber dates from the Cretaceous, perhaps dinosaurs were bitten, though there were other animals around at the time that the insects could have fed on.

Apart from the insects, other animals or their parts have been preserved in Burmese amber – these include bird feathers, lizard remains, worms, snails, velvet worms, scorpions, spiders, pseudoscorpions, mites, ticks, centipedes and millipedes. Interestingly, the amber sometimes contains conical tubes that have their small end at the surface of the amber, then widen into the amber where they finish in a rounded end. These were originally thought to be fungus fruiting bodies but are now known to be the burrows of bivalves, which bored into the amber after it had hardened and lay in water. Interesting recent research has been done on the microbes in Burmese amber. Some show a remarkable resemblance to ones that are known to spread disease today. The guts of a sandfly and a biting midge in Burmese amber have revealed blood, and within the blood *Leishmania* and malaria-like microbes

have been found. It has been suggested that these diseases may have contributed to the extinction of the dinosaurs.

ABOVE (fig 51) Female digger wasp (Hymenoptera: Sphecidae) in Burmese amber, one of nine in the same piece of amber demonstrating social behaviour. (Length 2.3 mm, or ¹⁄₁₆ in.)

DOMINICAN AMBER

Dominican amber comes from the Dominican Republic on the island of Hispaniola in the Greater Antilles (Caribbean, West Indies), not the island of Dominica in the Lesser Antilles. Although this amber was first mentioned by Christopher Columbus, interest in it only really started in 1960. There are many amber mines on the island, most of which occur in a mountain range called the Cordillera Septentrional. They consist

(fig 52) Piece of Dominican copal. Although it looks very similar to amber it can be distinguished using the alcohol test. (Length 53 mm, or 2 1/16 in.)

(fig 53) Piece of blue Dominican amber. (Length 85 mm, or 3 3/8 in.)

of pits or tunnels that are mined by hand by the local inhabitants. Occasionally, the tunnels fill with water and collapse. The amber is taken to the town of Santiago, then to Santo Domingo, where it is checked by the amber museum before being sold by dealers. Unfortunately, by the time it reaches the open market the information on which mine it came from is lost. Suggestions as to its age have ranged from Late Eocene to Middle Miocene, but most of the insect-bearing amber was deposited during the Early to Middle Miocene. A couple of mines produce copal of much younger age (fig 52), although this does contain some extinct species.

Dominican amber is commonly sold commercially both as jewellery and as pieces with inclusions. It is usually perfectly clear and comes in a variety of colours. Many pieces show a range of shades of yellow and orange, formed by a series of flows of slightly different colours (fig 54). It can also be green and blue (fig 53), although both these colours tend to fade with time. Dominican amber can contain cracks, air bubbles and water droplets. Only rarely is it cloudy, or contains pyritized cracks, or stellate (star-shaped) hairs. The insects are more diverse and generally better preserved than those in Baltic amber. There are many tropical forms present, indicating that the amber was produced in a tropical forest. So far, the insect fauna has not been as well studied as that from Baltic amber. Dominican amber is chemically similar to East African copal and was produced by the same type of legume tree belonging to the genus *Hymenaea*. Plant fossils in the amber indicate the tree that produced Dominican amber differed from species living today and it has been named *Hymenaea protera*.

(fig 54) Piece of Dominican amber containing a cockroach and two termites. The cockroach is on the left and the termites are close together on the right. (Length 53 mm, or 2 1/16 in.)

UK (ISLE OF WIGHT AND HASTINGS)

Isle of Wight amber is of particular interest because it is from the Early Cretaceous, which makes it one of the oldest insect-bearing ambers in the world (about 130 million years old). It is generally brown and cloudy with transparent yellow swirls, a lot of plant debris and some pyrite crystals (fig 55). A few similar pieces of amber have been found at Hastings, East Sussex (fig 56), and recently, spider silk was found in it. At about 140 million years old, this Hastings amber is slightly older than that from the Isle of Wight.

BELOW LEFT (fig 55) Spider in Isle of Wight, UK amber. (Length of longest leg 4.5 mm, or ³⁄₁₆ in.)

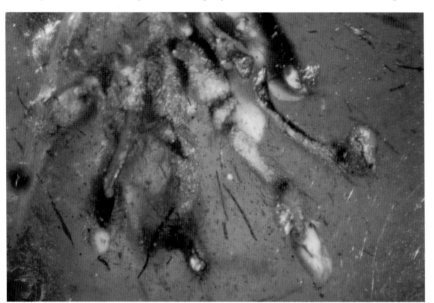

LEFT (fig 56) Amber from Hastings, East Sussex, UK. (Width 33 mm, or 1 ⁵⁄₁₆ in.)

RIGHT (fig 57) A piece of amber from Alava, Spain. (Length 100 mm, or 3 $^{15}/_{16}$ in.)

SPAIN

Spanish amber was first studied in 1762, though the importance of amber from this country was not realized until the 1990s when a Early Cretaceous insect-bearing amber deposit was discovered in Alava, Basque Country. This amber ranges from yellow to red, though it is usually orange and it can be transparent or cloudy (fig 57). The pieces are generally small and brittle, though the largest piece found is 200 mm, or about 7$^{7}/_{8}$ in, long. This amber has yielded 12 orders and many new species of insects, as well as other inclusions, including rare feathers. It was probably produced by an araucarian pine.

FRANCE

There are over 70 amber localities recorded in France, though most of the records are old. The most important insect-bearing ones are from the Aquitaine Basin in the southwest of the country (Cretaceous, about 110–95 million years old, fig 58), and from Oise in the north (Early Eocene, about 53 million years old). Of the former, an

RIGHT (fig 58) Several pieces of amber from the Aquitaine Basin, France.

important deposit was discovered at Archingeay, which yielded 60 kg (132 lb) of amber in 1999, and 15 orders of insects have been recorded from it so far. A new technique, called phase contrast X-ray synchotron imaging, has been used successfully to study cloudy amber from La Buzinie and produce three-dimensional images of insects that are not normally visible (fig 59). The younger Oise amber was only discovered in 1996, yet 20,000 pieces of amber have now been collected, containing 17 orders of insects. This amber was produced by a flowering tree, unlike the slightly younger Baltic amber.

MEXICO

Mexican amber is generally yellow with a brownish tinge, although red, green and blue forms also occur (fig 60). This amber often has parallel cracks running through it and this is a good indicator of genuine Mexican amber. It is mined in Chiapas State and is Late Oligocene to Early Miocene in age.

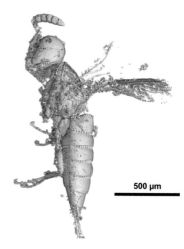

500 μm

ABOVE (fig 59) Three-dimensional image of a wasp produced from cloudy amber from La Buzinie, France. (Length 1.5 mm, or 1/16 in.)

LEFT (fig 60) Piece of Mexican amber. (Length 45 mm, or 1 3/4 in.)

LEBANON

Lebanese amber is generally yellow. It contains a lot of cracks, which make it very fragile (fig 61). Similar amber has been found in Israel and Jordan. Lebanese amber is important because it is Early Cretaceous in age and thus one of the oldest ambers to contain insects.

LEFT (fig 61) Piece of Lebanese amber. (Length 21 mm, or 15/16 in.)

RIGHT (fig 62) A piece of Sicilian amber, containing two spiders. (Length 40 mm, or 1 9/16 in.)

BOTTOM RIGHT (fig 63) A piece of very dark Sicilian amber. (Length 40 mm, or 1 9/16 in.)

SICILY

Sicilian amber (called simetite) is usually orange or red, but can also be green, blue and black (figs 62, 63). It is usually transparent but can also be cloudy. When polished, the surface is highly reflective, almost dazzling. It is Oligocene in age.

CHINA

Chinese amber is usually transparent and is orange or red in colour. Large pieces have been ornately carved (fig 64), but the source of these is unknown; they may have originally come from Burma (Myanmar). Amber with insects has been collected from Fushun in Liaoning Province and is Eocene in age.

BORNEO

Amber from Borneo is usually very dark red, almost black (fig 65), although yellow pieces also occur. The amber is generally slightly cloudy so any inclusions appear blurred. Some of the yellow pieces have not fully fossilized and are still copal. Borneo amber is Middle Miocene in age.

ABOVE (fig 65) A piece of Borneo amber. (Length 53 mm, or 2 ¹⁄₁₆ in.)

BOTTOM (fig 64) A carved piece of Chinese amber. (Length 120 mm, or 4 ¾ in.)

CHAPTER 3
Amber inclusions

INCLUSIONS ARE ALL THE OBJECTS THAT ARE TRAPPED IN AMBER. The most familiar of these are insects and spiders, but there are many other things that have also been trapped. These include bacteria, fungi and many different types of plants. Invertebrate animals (those without backbones), other than insects and spiders, include worms (fig 67), snails (fig 68), rotifers, tardigrades (water bears) and microscopic protozoa (see fig 87, p.44). You can also find vertebrates, but these are extremely rare and consist of frogs, lizards (fig 66), birds' feathers and mammal remains. The inclusions are very important for studying past diversity, ecology and biogeographical distributions.

There is a bias in the size of the animals that are trapped in amber. Animals that are larger than 20 mm (or ¹³/₁₆ in) long, such as lizards, frogs, scorpions, large insects and spiders, are generally strong enough to pull themselves free of the sticky resin, so the majority of inclusions that are trapped are much smaller – only a few millimetres long.

OPPOSITE (fig 66) Lizard in Dominican amber. (Length 55 mm, or 2 ³/₁₆ in.)

BELOW LEFT (fig 67) Roundworm (nematode) in Baltic amber. (Length 2.6 mm, or ¹/₈ in.)

BELOW (fig 68) Snail in Burmese amber. The small black inclusions are its droppings. (Length of piece 12 mm, or ½ in.)

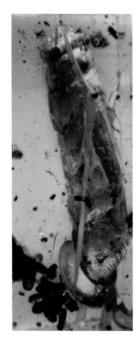

For example, the largest natural inclusions in amber in the collections at the Natural History Museum, London, are 20 mm (or ¹³⁄₁₆ in) long (fig 69) and the smallest known is 0.25 mm long (see fig 170, p.100).

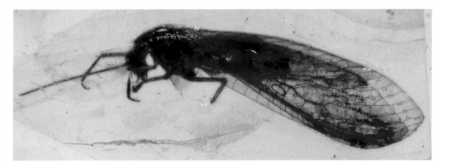

see fig 170, p.100

EVIDENCE OF ANIMAL BEHAVIOUR

Amber can preserve evidence of the behaviour of animals before and after they are trapped. When an animal gets trapped in resin it is still alive, so there is sometimes evidence of a struggle, such as concentric lines in the amber around it (fig 70). Some flies and harvestmen are able to break off their legs to enable their escape. Isolated legs are often seen in amber, as well as flies with some of their legs lying nearby (fig 71). Some flies are trapped while still mating (fig 72), whereas some insects lay eggs in amber just before dying, and there are even a few examples of insect pupae in amber with the adult insects emerging from them. Other insects are incomplete and their struggle on the surface of the resin may have brought them to the attention of larger animals looking for a meal. A few inclusions in amber are mouldy and the mould probably formed after the animal died on the surface of the resin, before it became completely engulfed (fig 73). Specimens where there is direct evidence of two types of animal interacting are of particular interest, and will be discussed in more detail later (see p.43).

see p.43

RIGHT (fig 69) Alder fly in Baltic amber. (Length 21 mm, or ¹⁵⁄₁₆ in.)

BELOW (fig 70) Planthopper bug in Baltic amber, with concentric lines around it, showing where it struggled. (Length 7 mm, or ¼ in.)

ABOVE (fig 71) Crane-fly in Baltic amber, which has broken off its legs in an effort to escape. (Length 3.5 mm, or ⅛ in.)

RIGHT (fig 72) Caught in the act – a pair of mating scavenger flies in Dominican amber. (Length of large fly 1.8 mm, or ¹⁄₁₆ in.)

BELOW (fig 73) Mouldy scuttle fly in Dominican amber. (Length including mould 3.6 mm, or ⅛ in.)

WHAT IS A SPECIES?

Many scientists are involved in work to describe and name the animals and plants in the world, both living and fossil. Each animal and plant is given a genus and species name in Latin. This system is known as binomial nomenclature and was established by Carolus Linnaeus, also known as Karl von Linné, in 1758.

A species is generally defined as a group of organisms that can only reproduce with each other. These organisms are alike and have a similar appearance, morphology, although you can get differences between the sexes, which is known as sexual dimorphism.

TAXONOMY

The process of describing and naming genera and species is known as taxonomy. A specimen that is used for the basis of a species description is known as the type. The different species are placed in a classification that consists of a hierarchy of taxonomic units, known as taxa (see example on the right). However, only a few of the taxa are used regularly. The most common units are: phylum, class, order, family, genus and species.

EVOLUTION AND EXTINCTION

A population is a large number of individuals of the same species living in a given area. The populations of a species define its biogeographical distribution. Populations can often split up or merge due to changes in the environment. If a population splits and the subsequent populations are then separated for some time, subtle changes can occur over the generations, and evolution takes place. If the populations are separated for long enough, they may change so significantly that if the populations merge they can no longer reproduce with each other. This means a new species has evolved, a process known as speciation.

Environmental changes put pressures on a species. It copes with these either by adapting and evolving or migrating to a more favourable area. If the environmental changes are rapid then the species may not be able to cope, in which case it will die out and become extinct. Some species adapt to changes better than others and some groups evolve faster than others. Species that have strict ecological requirements are more prone to extinction than those that do not. Species that are still living today are described as extant.

Though many species have been described and named, there are still many more that are waiting to be discovered. Insects have a high diversity and over a million extant species of insect have been described and named, but it is estimated that more than 10 million species are living on this planet today. Only about 30,000 fossil (extinct) insect species have been described and named but many more (estimates suggest 100,000) await description.

Classification of the honey bee

Phylum	Arthropoda
Superclass	Hexapoda
Class	Insecta
Subclass	Pterygota
Cohort	Neoptera
Superorder	Oligoneoptera (=Endopterygota)
Order	Hymenoptera
Suborder	Apocrita (Aculeata)
Superfamily	Apoidea
Family	Apidae
Subfamily	Apinae
Tribe	Apini
Genus	*Apis*
Subgenus	*Apis*
Species	*mellifera*
Subspecies	*mellifera*

RIGHT (fig 74) Honey bee, *Apis mellifera*, in East African copal. (Length 11.5 mm, or 7/16 in.)

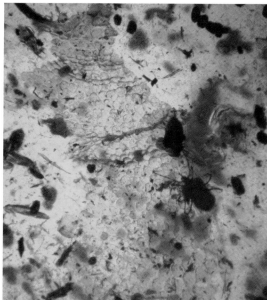

ABOVE (fig 75) A shed caterpillar skin in Dominican amber. (Length 3.4 mm, or ⅛ in.)

ABOVE RIGHT (fig 76) Piece of shed lizard or snake skin in Burmese amber with a mite lying over it. (Length of mite 1.5 mm, or ¹⁄₁₆ in.)

RIGHT (fig 77) A shed earwig skin in Baltic amber. This specimen has very long pincers. (Length 9 mm, or ⅜ in.)

BELOW (fig 78) Insect droppings in Dominican amber. (Length of topmost dropping 1.9 mm, or ¹⁄₁₆ in.)

ANIMAL PRODUCTS

The products of insects and other organisms are also trapped in amber. Insect droppings are very common in amber (particularly Dominican) and are small, black and barrel-shaped (fig 78). Sawdust plugs also occur; these would have been pushed out of holes by wood-boring beetles. Strands of silk from a spider's web are seen occasionally and even spiders' nests are known. When an insect or spider grows, it has to moult and shed its skin: often the skins have been preserved (figs 75, 77). Pieces of shed lizard or snake skin have also been found (fig 76). Ants and termites throw a lot of debris out of their nests, and this can become trapped in amber, although this

is not easily distinguishable from general debris. Ants eat other insects so their waste also includes lots of bits of insects. A vertebrate dropping or regurgitated pellet has been found in Baltic amber (fig 79). It consists of a concentration of beetle bits glued together. No one is sure what produced it – whether it came from a bird, lizard or even a bat.

PRESERVATION

The chemical properties of the different types of amber preserve the organisms in different ways. Dominican amber is the best preservative, with most insects appearing in perfect condition, including their internal tissues. Many Baltic amber insects are not as well preserved and often have a white coating around them. Most are also hollow due to decay, with little or no tissue preserved, while others are filled with pyrite crystals, which can also penetrate the wing membranes, turning them black. The insects in amber from Burma, Mexico and Borneo are generally less well preserved. Often they are semi-transparent, probably due to complete penetration by the resin, and may be partially dissolved, as many are incomplete and distorted (fig 80).

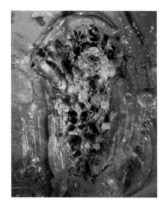

ABOVE (fig 79) Vertebrate dropping or regurgitated pellet in Baltic amber, consisting of beetle remains. It could have been produced by a bird, lizard or bat. (Length 6 mm, or ¼ in.)

BELOW (fig 80) Distorted planthopper bug in Mexican amber. (Length excluding wings 2.7 mm, or ⅛ in.)

RIGHT (fig 81) Air bubbles inside water drops in Baltic amber. (Length of longest drop 0.9 mm, or 1/32 in.)

RIGHT (fig 81) Air bubbles inside water drops in Baltic amber. (Length of longest drop 0.9 mm, or 1/32 in.)

INORGANIC INCLUSIONS

It is not only animals and plants that get trapped in amber. Bubbles and water droplets are also common. Occasionally, you can see water droplets that also have a bubble of air inside them. When the amber is turned, the bubble moves so that it stays at the top, acting as a natural spirit level (fig 81). The proportions of gases in the atmosphere have changed through geological time, and some scientists believe that amber bubbles can be used to study ancient atmospheres. However, there is a debate as to whether amber is a perfect sealant, as some scientists think that small molecules can migrate through amber and would therefore alter the proportions of gases in the bubbles. It is also likely that oxygen in the bubbles would have reacted with the amber. It was observed that all the water left the amber in fig 81 within five years after the photo was taken, so certainly the composition of these bubbles has changed and therefore does not accurately represent the atmosphere from long ago. Apart from the inclusions, flow planes and cracks also occur in amber. The flow planes indicate successive periods of resin flowing over each other. Sometimes they are crazed because the surface of the resin hardened before the next flow covered it (fig 82). The cracks would have developed after the amber formed, probably due to pressure generated by the weight of overlying sediments or tectonic movements (or as a result of the mining and polishing processes).

BELOW (fig 82) Crazed flow plane where the surface dried out prior to the next flow.

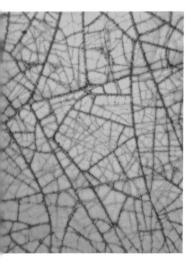

ANIMAL INTERACTIONS

Some amber specimens show evidence of behaviour that cannot otherwise be seen in the fossil record. Of particular interest is where a piece of amber contains direct evidence of two or more animals interacting with each other, either still attached or in close proximity. There are four types: parasitism, mutualism, commensalism and predation.

PARASITISM

Parasitism is where one organism benefits but the other (the host) suffers. By looking at their living relatives we can infer that many insects in amber would have been parasitic, for example parasitic wasps. The winged adult lays its eggs in the host, which can be an adult insect (or spider), pupa, larva or egg. The wasp larva (the parasite) hatches out and slowly devours its living host. Although there are many different species of wasp in amber, usually there is no way of knowing which other organisms they parasitized. The only way of knowing for sure is when you get a specimen with the parasite still attached to its host. There are rare examples of bugs in Dominican amber that have a sac attached to them with a developing wasp larva inside (fig 84). Occasionally you can get an indication of what animal an organism parasitized when you find several associated insects in one piece of amber. There are clear examples in amber of parasitic mites still attached to flies, caddis-flies and moths (fig 83). The earliest record of parasitism is in Lebanese amber. Some nematode worms are parasites inside insects. There are specimens of flies and other insects that have these worms coming out of their abdomens or lying in the amber nearby (see fig 67, p.37).

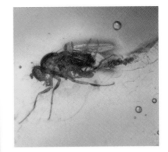

ABOVE (fig 83) Fruit fly in Dominican amber with a parasitic mite attached to its abdomen. (Length of fly 2.9 mm, or ⅛ in.)

BELOW (fig 84) Leafhopper bug in Dominican amber with a dryinid wasp sac attached to its head. (Length of bug excluding legs 2.9 mm, or ⅛ in.)

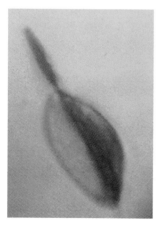

MUTUALISM

Mutualism is a partnership where each organism benefits from the other. Recently an example of this was discovered in Burmese amber of a winged termite (fig 86) with a damaged abdomen, with associated protozoa lying inside the wall of its gut and nearby in the amber (fig 87). These protozoa and bacteria would have helped the termite digest cellulose. Methane is produced as a by-product of this process and many termites in amber have large bubbles of methane projecting from their abdomens, which would have been produced after the termites were trapped. It is also possible that some methane was produced by the decay of the termites' internal tissues.

COMMENSALISM

Commensalism is where one organism benefits but the other is unaffected. One type of this is phoresy, where an animal hitches a lift on another. There are several examples of this in amber, particularly involving flightless arachnids getting rides from flying insects. Pseudoscorpions can be seen using their pincers to hang on to the legs and bodies of flies and wasps in Baltic amber (fig 85), and on to beetles in Dominican amber. Phoretic mites can be seen on flies, bees, beetles and termites in Dominican amber, although it is often difficult to tell whether mites that are associated with an insect are phoretic or parasitic. In Dominican amber, juvenile nematode worms have been recorded on ants and beetles, and a beetle larva has been recorded attached to the head of a bee.

PREDATION

Predation is where one animal eats another. Much of the evidence for this type of interaction is inferred because the type of insect observed in amber is wholly predatory today. Some insects in amber have been observed with other insects in their jaws. This can be due to one catching the other to eat just before entrapment, but instead it may be using its jaws to grab hold of something to try to pull itself free of the sticky resin (fig 88). Ants have been found with scale insects in their jaws that

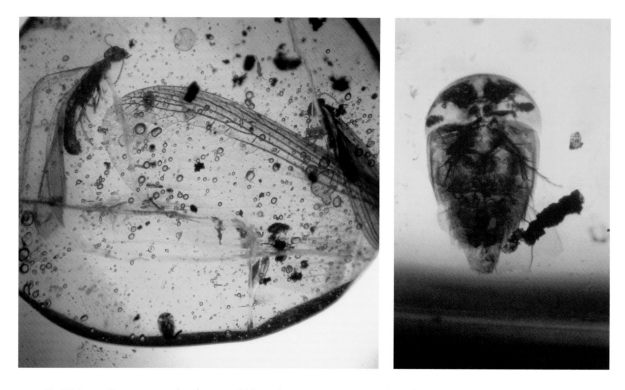

ABOVE (fig 89) Piece of Dominican amber containing termites, a flightless scuttle fly and a rove beetle. The fly and beetle would have lived inside termite nests. (Length of piece 20 mm, or ¹³⁄₁₆ in.)

ABOVE RIGHT (fig 90) Close-up of the rove beetle in fig 89. (Length 1.3 mm, or ¹⁄₁₆ in.)

BELOW (fig 91) Close-up of the flightless scuttle fly in fig 89. (Length 1 mm, or ¹⁄₃₂ in.)

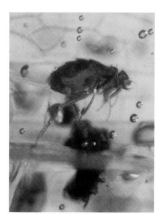

they would have been moving to another plant. Flightless female scuttle flies have been found in Dominican amber. Today, they are mainly scavengers in termites' and ants' nests, but some are specialized predators. The female crawls down into a nest, lays its eggs and crawls out. To evade detection she secretes the same chemicals that the ants and termites use to recognize each other. If it is a predatory species, then when the larvae hatch out they devour the termites, ants or their larvae. One scuttle fly occurs in a piece of Dominican amber in association with two species of termite and a rove beetle, which also lives in termite nests (figs 89, 90, 91). It is therefore very likely that this species of scuttle fly lived in the nests of termites rather than those of ants.

PLANT INCLUSIONS

The remains of plants are common in amber, although identifiable specimens of plant structures, such as leaves, twigs, cones and flowers are rare. The most common remains are fragments of bark and hairs from oak flowers (in Baltic amber). The bark fragments probably came from the trees that produced the amber. Pollen and spores also occur but they can only be seen by using a high-powered microscope. The Baltic amber flora has been well studied over the years, but little has been described from Dominican amber. Plant fossils in amber include mosses, liverworts, lichens, ferns, gymnosperms (including pines) and angiosperms (the flowering plants). In Baltic amber, most of the identifiable plant remains are gymnosperms or angiosperms, whereas the majority of specimens in Dominican amber are angiosperms.

GYMNOSPERMS

The gymnosperm specimens in Baltic amber are either from conifers or cycads. The conifers are the most common and consist of the twigs and cones of cypresses (fig 92), pines (fig 93), redwoods and podocarps. It is likely that some of the pine specimens came from the tree that produced Baltic amber, although the species (*Pinus succinifera*) was described based on the microscopic examination of bark.

BELOW LEFT (fig 92) Cypress twig, *Thuja*, in Baltic amber. (Length 13 mm, or ½ in.)

BELOW RIGHT (fig 93) Pine cone in Baltic amber. (Length 18 mm, or ¹¹⁄₁₆ in.)

ANGIOSPERMS

The angiosperm (flowering plant) remains consist mainly of leaves (fig 94) and flowers (fig 96), and over 60 families have been recorded. Most angiosperm specimens are difficult to identify by a non-expert; however, in Baltic amber there is one kind that is fairly common and easy to identify. Baltic amber commonly contains clusters of tiny stellate (star-like) hairs, which came from the male flowers of oaks. Not only are

BELOW (fig 94) Angiosperm leaf, *Eudaphniphyllum*, in Baltic amber. (Length 28 mm, or 1 ⅛ in.)

RIGHT (fig 95) Angiosperm flower, *Croton*, in Dominican amber. (Length 6 mm, or ¼ in.)

BELOW (fig 96) Bract, probably from the male flower of an oak tree, in Baltic amber. (Length 9 mm, or ⅜ in.)

the hairs preserved but so are isolated bracts (fig 95) and, rarely, complete flowers (fig 97) with the hairs still attached. The high abundance of oak hairs in Baltic amber indicates that most of the transparent resin was secreted during the spring and summer while the oaks were flowering.

Other trees that would have grown in the Baltic amber forest include maples, holly, beeches, chestnuts, laurels, magnolias, proteas and willows. Trees represented only by pollen include birch, horse chestnut and lime. Mistletoe would have grown on the trees.

Herbaceous and shrubby plants in Baltic amber include members of the palm, heather, *Geranium* and saxifrage families. In addition, there would have been grasses, represented by two species, though grasslands had not yet appeared. Other familiar flowering plants in the Baltic amber forest belong to the arum, lily, flax, olive, elm, rose, rock rose, tea, carrot and nettle families, although each family is only represented by one species.

In Dominican amber, leaves (fig 98) and flowers have been found belonging to the legume *Hymenaea*, which is the tree that would have produced the amber. The Dominican amber tree, which differs from living species and is therefore extinct, was named *Hymenaea protera*.

LEFT (fig 97) Male oak tree flower (*Quercus*) in Baltic amber. (Length 3.7 mm, or ⅛ in.)

THE ANCIENT AMBER FORESTS

The insects and other inclusions in amber can tell us a lot about the ecology of the ancient forests, a study known as palaeoecology.

The palaeoecology of the Baltic and Dominican amber-producing forests is better known than for any others. From the evidence available one can imagine what the forests were like. The subtropical Baltic forest would have been a mixture of coniferous and deciduous trees. The pine trees would have secreted copious amounts of resin as flows running down the trunks, stalactites dripping from branches and hardened lumps stuck to the bark. Oaks would have been flowering, with hairs and pollen from the flowers drifting in the warm breeze. You can imagine constant activity from the insects and other animals in the forest: caterpillars, crickets and stick insects munching on the leaves. Male crickets singing to attract mates. Columns of aphids (greenfly) on the twigs, each with its long proboscis inserted through the bark to suck the sap, constantly attended by ants that feed on their honeydew secretions and protect them from lacewings. Lines of ants marched up and down the trees in the forest.

The forest was likely to have been fairly dense, with clearings where old trees had died and rotted while other plants grew around them, competing for the sunlight. The many scavengers such as woodlice, earwigs, beetles, termites, ants, cockroaches, mites, barklice, millipedes, springtails, silverfish and bristletails would have crawled on the rotting logs, under the bark or in the leaf litter on the forest floor, munching on the decaying matter. Beetle grubs bored into the rotting wood to make a network of galleries, while female ichneumon wasps walked jerkily on the surface of the tree trunks, with their antennae waving, listening out for movement so that they could insert their ovipositors into the wood in the right place to lay their eggs in the grubs.

BELOW (fig 98) Leaf of *Hymenaea protera*, the tree that produced Dominican amber. (Length 52 mm, or 2 in.)

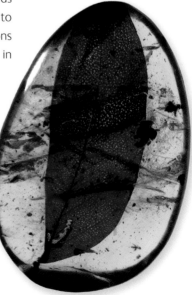

Mushrooms and other fungi growing on the logs and forest floor would have provided food for the larvae of fungus gnats. The gnats and midges would have formed mating swarms, and hoverflies would have flitted in and out of the sunbeams. Bees and wasps would have been buzzing around the flowers to pollinate them. Spiders would lie in wait in webs for their next struggling meal. Meanwhile, other predators such as centipedes, damselflies, harvestmen, praying mantises and various large flies actively hunted for their next meal or lay in wait for it to come along.

The vertebrates, perhaps mammals, birds, lizards and frogs, would have fed on the various insects and plants, and in turn provided food for biting and bloodsucking insects and parasites including mosquitoes, black-flies, biting midges, horseflies, sandflies, fleas and lice. There would have been occasional ponds, fed by streams in the forest that supported the predatory nymphs of damselflies, which may have fed on wriggling midge larvae. Also in the ponds and streams were scavenging mayfly and stonefly nymphs, and caddis-fly larvae that protected themselves in portable houses of plant fragments. On some days in the spring and summer, mayflies would have hatched, swarmed above the ponds, laid their eggs and died.

The animal and plant species in the Dominican amber forest were different, but the groups would have lived in a similar way to those in the Baltic forest. The essential difference was that the Dominican forest was tropical and therefore hotter, more humid and supported a greater diversity of species. The majority of plants would have been angiosperms, some of which would have been epiphytes (including bromeliads) that lived in the canopy. Some insect groups – the ants and termites – were probably more abundant in the Dominican forest than in the Baltic forest, whereas others, such as greenfly, were much rarer.

THE SEARCH FOR DNA

Deoxyribonucleic acid (DNA) is the molecule within cells that contains the information necessary for the growth and function of all living organisms. It is made out of subunits called nucleotides, which are joined together to form a double helix (fig 99). Biologists have been extracting it from organisms to investigate the relationships between them. In the 1980s it was realized that cell structures are preserved in insects in amber, so then the search was on to find DNA in the insects' cells.

BELOW (fig 99) The double helix structure of a DNA molecule.

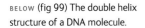

REPORTS OF DNA IN AMBER

DNA was first reported to have been recovered from amber in 1992 when scientists in California claimed to have extracted fragments of DNA from an extinct species of bee, *Proplebeia dominicana*, in Dominican amber. Shortly afterwards reports appeared of DNA extracted from an extinct species of termite, *Mastotermes electrodominicus*, (see fig 107, p.55), also in Dominican amber, by scientists in New York. This was followed by reports of DNA extraction from a beetle in Lebanese amber. However, only small bits of the DNA string were recovered. There has been some scepticism as to whether these claims are genuine or the result of contamination. Experiments on the survival rate of DNA have shown that it breaks down very quickly, particularly in the presence of water. However, the insects in amber are dehydrated, and if dehydration happened quickly then it could possibly halt the decay of the DNA.

At the Natural History Museum, London, scientists tried to repeat the experiments to obtain DNA from the Dominican amber bee, *Proplebeia dominicana* (fig 100). These bees are common in Dominican amber because they collect resin to make their nests. Several suitable specimens were selected, broken up and tested, but no insect DNA was recovered. This casts doubt on the earlier reports because it appears that the experiments are not replicable, which is a fundamental requirement for reliable scientific results. Subsequent attempts by other scientists have not been successful either.

ABOVE (fig 100) A bee in Dominican amber, of the species *Proplebeia dominicana*, which was used to attempt DNA extraction. (Length 4 mm, or ³⁄₁₆ in.)

OPPOSITE (fig 103) Mating pair of biting midges in Baltic amber. (Length of male, top 1.6 mm, or 1/16 in.)

RIGHT (fig 101) Mosquito in Dominican amber. (Length excluding upward-pointing proboscis 3.7 mm, or 1/8 in.)

IS *JURASSIC PARK* POSSIBLE?

Even if DNA could be extracted from insects in amber, a real-life *Jurassic Park* is not possible. There are many reasons why such a venture will remain fiction. First, there are no known insect-bearing Jurassic ambers. Second, contrary to popular belief, mosquitoes (Diptera: Culicidae) (fig 101) are extremely rare in amber. There is one described from Canadian amber and one from Burmese amber. There are only a handful known in Baltic amber and a few tens of specimens in Dominican amber. There are, however, other biting insects, known from Mesozoic deposits that may have fed on dinosaurs. Black-flies (Diptera: Simuliidae) (see fig 105) are known from Middle Jurassic deposits and two have been found in Cretaceous amber. The oldest horsefly (Diptera: Tabanidae) (fig 102) was found preserved in limestone of Early Cretaceous age in Dorset, England, but none are known in Cretaceous amber. Biting midges (Diptera: Ceratopogonidae) (fig 103) have been found in Burmese, Canadian, Siberian, Lebanese, Spanish and French (Archingeay) amber but they can feed on the blood of many things, including other insects. A few are known in Canadian amber that have jaws adapted for biting vertebrates, but it is debatable whether they could have fed on dinosaurs. Sandflies (Diptera: Psychodidae) have been found in Burmese and Lebanese amber and one has large mouthparts similar to those of a living species

BELOW (fig 102) Horsefly in a piece of Baltic amber. (Length 15 mm, or 9/16 in.)

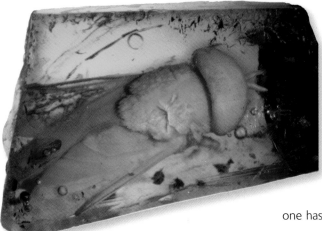

that feeds on the blood of crocodiles (fig 104). This type of fly could well have fed on the blood of dinosaurs. However, it is extremely unlikely that anyone will destroy this specimen on the remote chance of extracting DNA because it is a male, and only female sandflies feed on blood.

Even after an insect is trapped in resin, bacteria and enzymes continue working in the gut, rotting the insect from the inside. Indeed, many insects preserved in amber, particularly Baltic, are completely hollow without any internal tissue preserved. If it is so difficult to get DNA from an insect in amber, then the chances of getting any DNA from something it fed on are even more remote. If it were possible to extract DNA from a blood meal in an amber insect, only tiny amounts of the entire DNA string (genome) would be recovered and it would probably be contaminated with bacterial and insect DNA. Key parts of the genome would be required to work out which type of animal the blood came from. Biologists could only guess at what was missing from the complete DNA string. Although scientists can manipulate and make copies of DNA, they can't make it grow into an animal.

ABOVE (fig 104) Sandfly in Lebanese amber. This type of fly may have fed on the blood of dinosaurs! (Length 1.5 mm, or 1/16 in.)

RIGHT (fig 105) Black-fly in Baltic amber. This fly may belong to the subgenus *Morops*, which is only found today in Southeast Asia. (Length 2.1 mm, or 1/16 in.)

BIOGEOGRAPHY

The study of the geographical distribution of animals and plants is called biogeography. Many entomologists study present-day insect species and their biogeographical distributions to work out how they evolved from other closely related species and also where this speciation probably took place. Some scientists have used the theory of plate tectonics to explain the occurrence of closely related species in different parts of the world. However, the study of insects preserved in amber is yielding important information on past changes in biogeographical distributions that cannot be deduced solely from looking at the distributions of present-day species. Indeed, some occurrences in amber are completely unexpected because their living relatives are found on the other side of the world today. Clearly, plate tectonics has had little or no effect on these distributions. The distributions were probably controlled by environmental and/or ecological pressures, which made the species migrate from one area to another or made large populations shrink, as illustrated in the examples of flies and termites below.

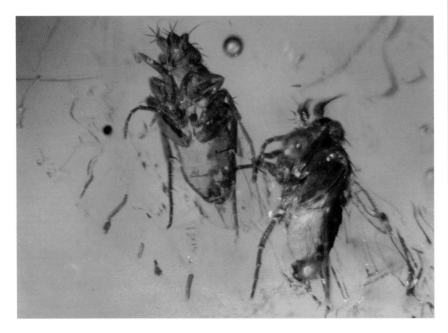

LEFT (fig 106) Scuttle flies (genus *Abaristophora*) in Dominican amber. This genus is only found today in Nepal and New Zealand. (Length of fly on right 1.4 mm, or ¹/₁₆ in.)

FLY DISTRIBUTIONS

There are many species of fly in Baltic amber whose nearest relatives live in Southeast Asia and/or southern Africa today – they no longer live in Europe. For instance, the black-fly in fig 105 may belong to the subgenus *Morops*, which is only found today in Southeast Asia. The snipe fly family Rachiceridae (see fig 85, p.44) lives mainly in Southeast Asia and North and South America today, with only one species living in central Spain. The closest relatives of the Baltic amber horsefly in fig 102 live today in Australia, southeast Africa, Chile and eastern USA. Similar surprising occurrences are being recognized in Dominican amber, such as scuttle flies belonging to the genus *Abaristophora* (fig 106). This genus only occurs today in Nepal and New Zealand.

TERMITE DISTRIBUTIONS

There is a primitive genus of termites called *Mastotermes* that lives only in Australia. This genus has turned up in Mexican and Dominican amber (fig 107), and many Eocene, Oligocene and Miocene fossil insect localities in Europe (including the Isle of Wight and Hampshire, England). So this termite must have had a worldwide distribution in the past, which has been reduced to a relict distribution in Australia (it has been accidentally introduced to New Guinea and New Zealand by humans).

BELOW (fig 107) Large termite in Dominican amber. This termite belongs to the genus *Mastotermes*, which is found today in Australia. (Length 26 mm, or 1 in)

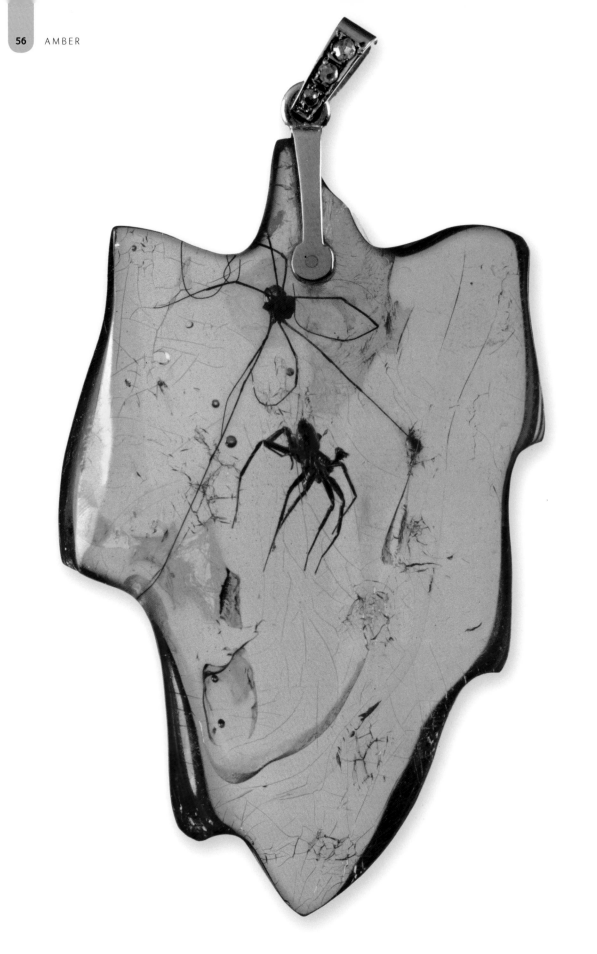

CHAPTER 4
Arthropods

T HE PHYLUM ARTHROPODA CONSISTS of all the invertebrates with jointed limbs and an exoskeleton generally made of chitin. The body is segmented and is usually divided into a head, thorax and abdomen, although in some groups the head and thorax are fused. They grow by moulting and often have compound eyes consisting of many lenses. This phylum contains five superclasses. Four are found in amber and are briefly described below; the fifth is the extinct Trilobitomorpha (trilobites).

CRUSTACEA – CRABS AND THEIR RELATIVES
The Crustacea have a head-shield (carapace) and two pairs of antennae. There are many different groups of predominantly sea-dwelling Crustacea, including crabs, lobsters, shrimps, prawns, crayfish and barnacles. Three orders belonging to the class Malacostraca have been recorded in amber: the Amphipoda, Isopoda and Decapoda. Amphipods are laterally compressed (=flat sides) and are curved. They usually live in the sea or in freshwater and some are known commonly as sandhoppers. They are extremely rare in amber but have been recorded in Baltic and Dominican amber. Isopods, by contrast, are dorso-ventrally flattened (=flat top) (fig 109). They live in the sea, freshwater or on land and are known commonly as woodlice or sea slaters. They are more common than the amphipods in amber but they are still very rare. Only one specimen of Decapoda has been recorded – a small crab (Suborder Brachyura) in Dominican amber that would have lived in bromeliads.

ABOVE (fig 109) Woodlouse (Crustacea: Isopoda) in Dominican amber. (Length 2 mm, or ¹⁄₁₆ in.)

CHELICERATA – SPIDERS AND THEIR RELATIVES
The Chelicerata have their head and thorax fused together to form a cephalothorax, which has six pairs of appendages and no antennae. The first pair of appendages at the front of the head often form curved, downward-pointing fangs known as chelicerae. The second pair are known as pedipalps, which are variable in appearance and have either a sensory, food-gathering or reproductive function. The other

OPPOSITE (fig 108) Baltic amber pendant containing a spider (Araneae) and harvestman (Opiliones). The harvestman (daddy-long-legs) is at the top. (Length of pendant excluding ring 50 mm, or 1 ¹⁵⁄₁₆ in.)

four pairs of appendages form the characteristic eight legs. This group contains two classes, the Arachnida (spiders, mites and scorpions) and the Merostomata (horseshoe crabs and extinct sea-scorpions), although only the Arachnida occur in amber. Arachnids are predators and feed on other invertebrates. Eight orders have been recorded in amber.

- The order Araneae (=Araneida) are the spiders with fang-like chelicerae and short, sometimes bulbous pedipalps. They usually have an unsegmented, fat, hairy abdomen and hairy legs (fig 108). Spiders are common in amber and many different species have been described.

- Members of the order Opiliones are commonly known as harvestmen or 'daddy-long-legs'. They appear similar to spiders, except the cephalothorax and abdomen are joined as one and they are not hairy (fig 108). They also have long pedipalps and very long legs. Harvestmen are very rare in amber.

- The order Acari (=Acarida) are the mites and ticks. The cephalothorax and abdomen are joined as one, they usually have short legs and they are extremely small (often less than 1 mm, or ¹⁄₃₂ in, long) (fig 110). They can be smooth or hairy. Some juvenile mites only have six legs. Acarids are fairly common in amber but are often overlooked because of their small size. Mites are more common in Baltic amber than Dominican amber.

- The order Scorpiones are the scorpions. Their pedipalps are modified to form pincers and they have a long tail with a sting (fig 111). Scorpions are extremely rare in amber, although they are often seen as fakes in plastic.

- The order Pseudoscorpionida are the pseudoscorpions. They superficially resemble

ABOVE (fig 110) Mite (Arachnida: Acari) in Baltic amber. (Length excluding legs 0.4 mm, or ¹⁄₅₀ in.)

RIGHT (fig 111) Scorpion (Arachnida: Scorpiones) tail in Burmese amber. (Length 8.5 mm, or ⁵⁄₁₆ in.)

scorpions in that they have pincer-like pedipalps; however, they differ from scorpions in that they do not have a tail and they are much smaller (only a couple of millimetres long) (fig 112). They also have a fat, segmented body and are very rare in amber.

- The order Amblypygi are commonly known as tail-less whip scorpions. They are large arachnids with long spiny pedipalps and the front pair of legs are considerably longer than the others and are used as feelers. They are extremely rare in amber.

- The Solpugida are commonly known as wind spiders. They are large and have four very large chelicerae; the two upper fangs point downwards and the two lower fangs point upwards. They act like pincers, with a similar action to a staple remover. The pedipalps are similar to the legs except they have flat adhesive pads at the end. Only one solpugid has been recorded from amber (Dominican).

- The Schizomida does not have a common name. Schizomids are small, blind arachnids (up to 6 mm, or ¼ in, long) with a segmented cephalothorax and a short tail (fig 113). Their chelicerae are large and fang-like. The front pair of legs are very long and used as feelers. They have only been found in Dominican amber.

ABOVE (fig 112) Pseudoscorpion (Arachnida: Pseudoscorpionida) in Baltic amber. (Length including claws 3.2 mm, or ⅛ in.)

LEFT (fig 113) Schizomid (Arachnida: Schizomida) in Dominican amber. (Length including front legs 7.5 mm, or ⁵⁄₁₆ in.)

MYRIAPODA – CENTIPEDES AND MILLIPEDES

Members of the superclass Myriapoda are long and thin with many body segments, many pairs of legs and a pair of antennae. There are four classes, all of which have been recorded from amber. The Chilopoda comprise the centipedes. They have a single pair of legs on each body segment and a pair of curved fangs situated under the head (fig 114). The Diplopoda comprise the millipedes. They have two pairs of legs on each body segment and generally have many more legs than centipedes (fig 116). They do not have fangs and, in contrast to centipedes, are vegetarians rather than carnivores. Members of the distinctive order Polyxenida are very hairy (fig 115). The Symphyla are similar to centipedes in that they have a single pair of legs on each segment, but they do not have fangs. The Pauropoda are very small, up to 2 mm ($\frac{1}{16}$ in) long with less than 12 pairs of legs – two pairs per body segment when looking from above. Myriapods are very rare in amber and the Pauropoda were only recently recorded in Baltic amber. Centipedes are more common in Baltic amber than Dominican amber, although the reverse is true for millipedes.

HEXAPODA – INSECTS

The arthropods in the superclass Hexapoda have six legs and include the insects. They are the most abundant and diverse group of arthropods and will be described in more detail later.

BELOW (fig 114) Centipede (Myriapoda: Chilopoda) in Baltic amber. (Width of centipede 1.1 mm, or $\frac{1}{16}$ in.)

BELOW RIGHT (fig 115) A very hairy millipede (Diplopoda: Polyxenida) in Baltic amber. (Length excluding hairs 4.5 mm, or $\frac{3}{16}$ in.)

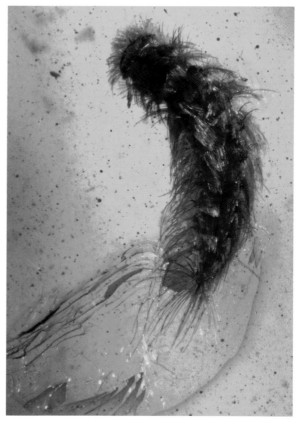

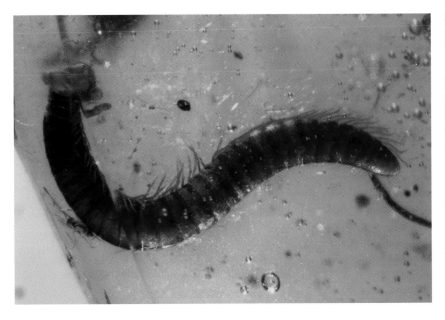

LEFT (fig 116) Millipede (Myriapoda: Diplopoda) in Dominican amber. (Width of millipede 0.6 mm, or ¹⁄₅₀ in.)

KEY TO ARTHROPODS

Keys to identification The keys in this book are designed to identify most of the arthropod inclusions that have been found in amber, and not living arthropods. They use the most obvious characteristics and are easy to follow. Many amber inclusions can only be examined from one side and parts of them are often obscured by cracks and bubbles. Pieces of amber may also be rounded, which can make the inclusion difficult to see due to distortion. If this is the case, you may need to look at the photographs and descriptions to aid identification.

The rarity of an animal is indicated after its name by: VC very common, C common, R rare, VR very rare, (–) not recorded. Where this is given twice and separated by a slash, this indicates rarity in Baltic and then Dominican amber. For example, C/VR means common in Baltic amber but very rare in Dominican amber.

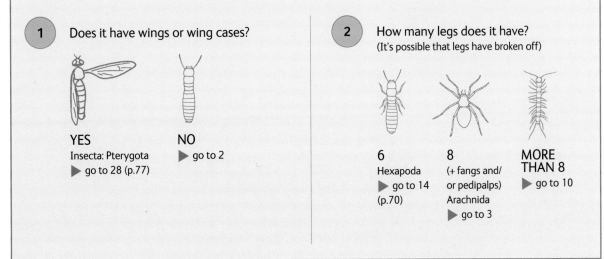

1 Does it have wings or wing cases?

YES
Insecta: Pterygota
▶ go to 28 (p.77)

NO
▶ go to 2

2 How many legs does it have?
(It's possible that legs have broken off)

6
Hexapoda
▶ go to 14
(p.70)

8
(+ fangs and/ or pedipalps)
Arachnida
▶ go to 3

MORE THAN 8
▶ go to 10

KEY TO ARTHROPODS

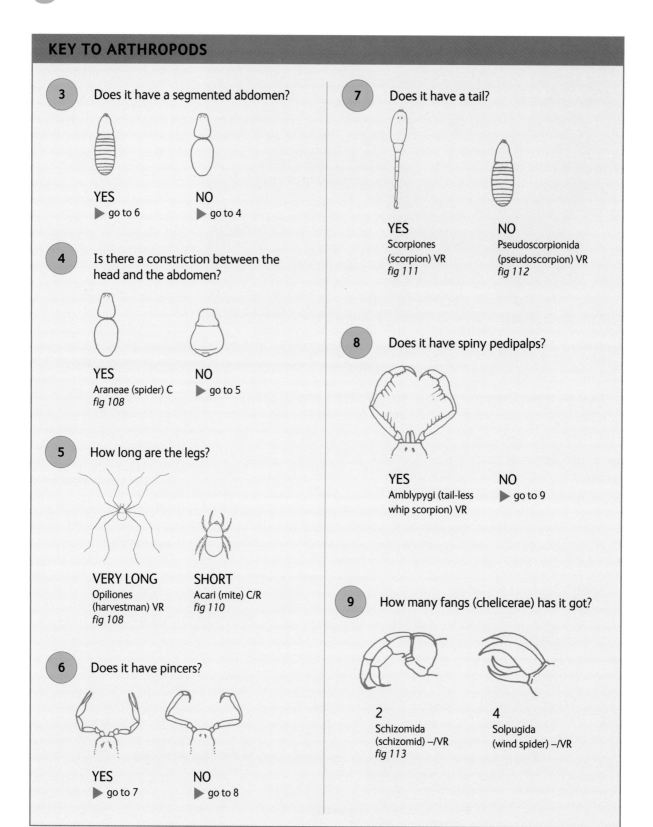

3 Does it have a segmented abdomen?

YES
▶ go to 6

NO
▶ go to 4

4 Is there a constriction between the head and the abdomen?

YES
Araneae (spider) C
fig 108

NO
▶ go to 5

5 How long are the legs?

VERY LONG
Opiliones
(harvestman) VR
fig 108

SHORT
Acari (mite) C/R
fig 110

6 Does it have pincers?

YES
▶ go to 7

NO
▶ go to 8

7 Does it have a tail?

YES
Scorpiones
(scorpion) VR
fig 111

NO
Pseudoscorpionida
(pseudoscorpion) VR
fig 112

8 Does it have spiny pedipalps?

YES
Amblypygi (tail-less
whip scorpion) VR

NO
▶ go to 9

9 How many fangs (chelicerae) has it got?

2
Schizomida
(schizomid) –/VR
fig 113

4
Solpugida
(wind spider) –/VR

KEY TO ARTHROPODS

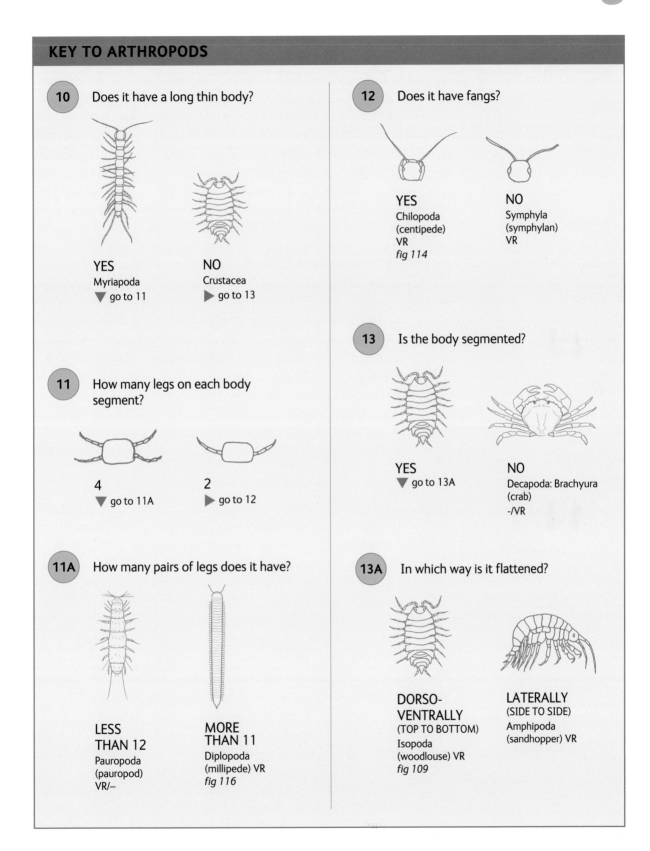

10 Does it have a long thin body?

YES
Myriapoda
▼ go to 11

NO
Crustacea
▶ go to 13

11 How many legs on each body segment?

4
▼ go to 11A

2
▶ go to 12

11A How many pairs of legs does it have?

LESS THAN 12
Pauropoda
(pauropod)
VR/–

MORE THAN 11
Diplopoda
(millipede) VR
fig 116

12 Does it have fangs?

YES
Chilopoda
(centipede)
VR
fig 114

NO
Symphyla
(symphylan)
VR

13 Is the body segmented?

YES
▼ go to 13A

NO
Decapoda: Brachyura
(crab)
-/VR

13A In which way is it flattened?

DORSO-VENTRALLY
(TOP TO BOTTOM)
Isopoda
(woodlouse) VR
fig 109

LATERALLY
(SIDE TO SIDE)
Amphipoda
(sandhopper) VR

THE INSECT FOSSIL RECORD

Amber is extremely important for documenting the fossil record of terrestrial animals, particularly small insects that usually do not get preserved in sedimentary rocks. The amber insect record is complemented by insect-bearing lake sediments that preserve the larger insects such as dragonflies (Odonata) and grasshoppers (Orthoptera), which do not get trapped in amber. New discoveries are regularly being made in amber; for example, some families of insects and other arthropods were discovered in the amber collection at the Natural History Museum, London, which had not previously been recorded in amber. Some of these were not even known to have a fossil record and a few were new to science (e.g. Corydasialidae, see fig 69, p.38).

THE BALTIC AMBER FAUNA

The best studied amber insect fauna is that from the Baltic. About 5,000 species have been described and named so far, but this is by no means all of them. The total Baltic insect fauna about 35 million years ago may have been about 10,000 species. Some groups, although well represented in the amber, have hardly been studied, such as some families of parasitic wasps. Studies on the Dominican amber fauna only started about 40 years ago and so far over 500 species have been described and named, and many more are currently being studied. The Dominican amber fauna is tropical and more diverse than that of the Baltic. Perhaps as many as 20,000 species were present in the Dominican Republic at the time. A database of fossil insect species is being developed; however, given that there are so many insect species that have not yet been named from amber or other fossil insect deposits, it is not yet reliable enough to determine the times of either the first appearance (i.e. origination) or the extinction of insect species.

The family is the most useful and reliable taxonomic unit to study insect origination and extinction. There are over 1,500 families of insect in the fossil record, compared to about 1,000 families living today, of which about 70% have a fossil record. Figure 117 illustrates the number of insect families at different periods of time, according to fossil evidence.

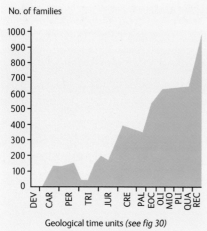

ABOVE (fig 117) Graph showing the number of families of insects through time, according to the fossil record. The apparent jump from the Quaternary (Qua) to Recent (Rec) is a reflection of the number of living families without a fossil record.

THE CRETACEOUS SURVIVORS

Amber has made the biggest contribution to the Late Cretaceous, Eocene, Oligocene and Miocene insect record. It is interesting to note that 90% of insect families recorded from Canadian and Siberian amber (Late Cretaceous) are represented by insects living today. This indicates that insects were hardly affected at the family level by the extinction at the end of the Cretaceous, which was when the dinosaurs became extinct. Although we do not know how many species went extinct, a study of fossil plant leaves across the boundary shows that there is much less insect damage at the beginning of the Paleocene, so insect populations had been adversely affected. A larger extinction of insect families took place during the Early Cretaceous

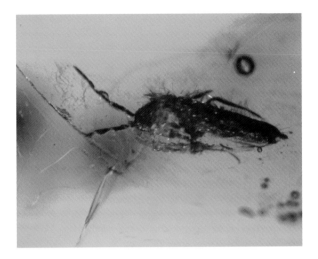 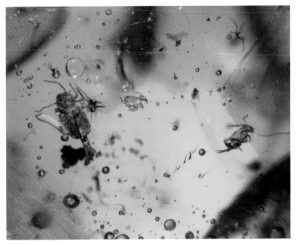

due to the radiation of the flowering plants, which changed terrestrial ecosystems. The earliest records of extant insects come from Baltic amber. Eight species have been recorded but re-examination of five of them has shown that they are fake (see fig 11, p.10), misidentified or preserved in copal. Of the three, one is a large, predominantly ground-dwelling beetle, so it is likely that this is also a fake. The other two records are as yet undisputed: they are a specimen of the mayfly *Heptagenia fuscogrisea* and several specimens of the tiny fairy fly *Palaeomymar duisburgi*, (see fig 172, p.102) which is actually a parasitic wasp. Both belong to primitive groups that have hardly changed with time. Some species alive today are also known from Dominican and Mexican amber.

INSECTS

The superclass Hexapoda comprises four classes, all the members of which possess a three-segment thorax that has a single pair of legs on each segment and an abdomen with 12 or fewer segments. The first three classes, of which two have been recorded in amber, are primitive wingless hexapods (apterygotes) that have internal (entognathous) mouthparts. The Protura have not been found in amber.

COLLEMBOLA – SPRINGTAILS

The class Collembola are commonly known as springtails because they have a forked spring attached to the tip of the abdomen. At rest it is folded under the body (fig 118) but if the collembolan is disturbed, it flicks the fork backwards, which propels the animal into the air. Springtails are small (only a couple of millimetres long) and are fairly rare in amber, although they are often overlooked because of their small size. Usually the forked spring can be seen projecting backwards, which is probably due to the animal trying to use it to escape from the amber. Springtails can be smooth or hairy and some have long antennae that curve outwards. One distinctive family recorded in amber is the Sminthuridae, which have a smooth, short, fat abdomen with a knob on the end (fig 119).

ABOVE LEFT (fig 118) Springtail (Collembola) in Dominican amber. This example has not released its spring, which is still folded under its body. (Length 1.3 mm, or 1/16 in.)

ABOVE (fig 119) Three springtails (Collembola: Sminthuridae) in Baltic amber. (Length of largest on the left, including spring 1.8 mm, or 1/16 in.)

DIPLURA – TWO-TAILED BRISTLETAILS

The class Diplura are commonly known as two-tailed bristletails because they have two segmented appendages (cerci) at the tip of the abdomen. Unlike springtails, they do not have eyes and are blind. Diplurans are extremely rare in amber.

INSECTA – TRUE INSECTS

The majority of hexapods belong to the class Insecta, which are characterized by having external (ectognathous) mouthparts. The Insecta are split into two subclasses.

APTERYGOTA – WINGLESS INSECTS

The subclass Apterygota are the primitive insects that do not have wings and comprise two extant orders – the Archaeognatha and Zygentoma (also known as Thysanura). Both orders are similar in that they have long antennae and at least three long tails (caudal filaments) at the tip of the abdomen. They differ in that the Archaeognatha (known commonly as bristletails) have large eyes (fig 120). The Zygentoma (known commonly as silverfish), by contrast, have small eyes (fig 121). Both orders are rare in amber.

ABOVE (fig 120) Bristletail (Archaeognatha) in Baltic amber. (Length including tail 15 mm, or ⁹⁄₁₆ in.)

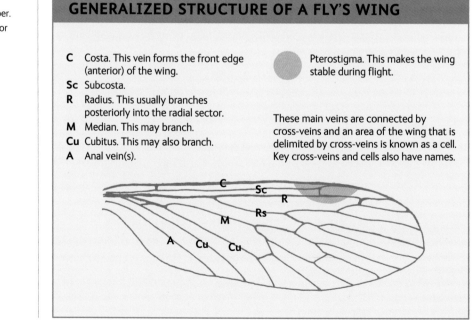

GENERALIZED STRUCTURE OF A FLY'S WING

C Costa. This vein forms the front edge (anterior) of the wing.

Sc Subcosta.

R Radius. This usually branches posteriorly into the radial sector.

M Median. This may branch.

Cu Cubitus. This may also branch.

A Anal vein(s).

Pterostigma. This makes the wing stable during flight.

These main veins are connected by cross-veins and an area of the wing that is delimited by cross-veins is known as a cell. Key cross-veins and cells also have names.

PTERYGOTA – WINGED INSECTS

The majority of insects belong to the subclass Pterygota, most of which possess two pairs of wings. The group also includes wingless groups (e.g. fleas) that evolved from winged ancestors. The wings have a network of veins that branch and are characteristic for each group of insect (see panel opposite). The wings are very useful for identifying the insects trapped in amber.

The Pterygota are divided into two smaller groupings – the Palaeoptera and Neoptera. The Palaeoptera comprise those insects that have outstretched wings whereas Neoptera are able to fold their wings back over their body.

PALAEOPTERA – OUTSTRETCHED WINGS

The Palaeoptera contains several orders, most of which are extinct. There are two living orders: the Odonata and Ephemeroptera. The Odonata are predators and have very short antennae, very large eyes, powerful jaws, a long abdomen and equal-sized wings with many veins, cross-veins and a pterostigma. They comprise the

BELOW (fig 121) Silverfish (Zygentoma) in Baltic amber. (Length 7 mm, or ¼ in.)

RIGHT (fig 122) A pair of
overlapping damselfly (Odonata:
Zygoptera) wings in Baltic amber.
(Length 17 mm, or ¹¹/₁₆ in.)

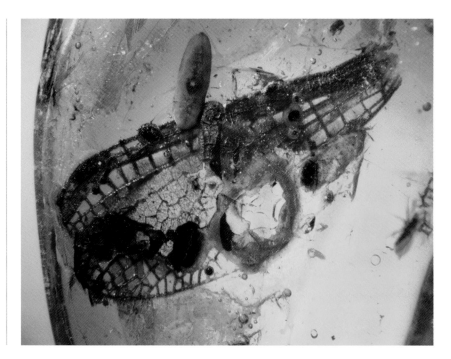

BELOW (fig 123) Mayfly
(Ephemeroptera) in Baltic amber.
(Length excluding legs and tail
6.5 mm, or ¼ in.)

dragonflies (Suborder Anisoptera) and damselflies (Suborder Zygoptera) (fig 122). The dragonflies are large and rest with their wings outstretched whereas damselflies are smaller and rest with their wings vertically. Dragonflies and damselflies are both extremely rare in amber, with dragonflies being rarer. The order Ephemeroptera comprise the mayflies. Mayflies have two or three long caudal filaments, short antennae and large compound eyes (see fig 50, p.29, fig 123). The wings usually have many cross-veins, do not possess a pterostigma and the hindwings are much smaller than the forewings. Many mayfly families have been recorded in amber, although they are rare. This is due to the fact that they do not live long (a couple of hours to several days) and do not venture far from their pond or river breeding grounds.

NEOPTERA — FOLDED WINGS

The majority of insects belong to the Neoptera, which consists of those insects that can fold their wings back along their body. The Neoptera can be divided into those that have incomplete metamorphosis (exopterygotes) and those that have complete metamorphosis (endopterygotes) (fig 124 below).

INCOMPLETE METAMORPHOSIS

Exopterygotes start life as an egg, which hatches into a nymph generally looking similar to the adult, except that it is wingless. The nymph progressively grows by moulting and produces wings at its last moult to become an adult. (Palaeoptera are also exopterygotes).

COMPLETE METAMORPHOSIS

Endopterygotes, by contrast, start life as an egg that hatches into a larva (also commonly known as a grub, maggot or caterpillar). The larva does not look anything like the adult and grows by progressively moulting. It then produces a pupa (chrysalis or cocoon) in which the tissue breaks down and re-forms (metamorphoses). The adult insect, complete with wings, then hatches out of the pupa.

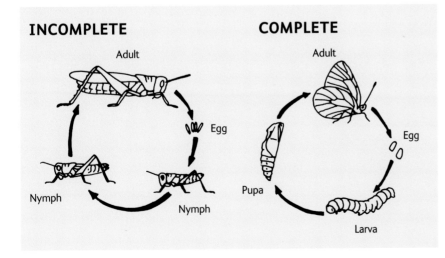

LEFT (fig 124) The differences between incomplete and complete metamorphosis.

KEY TO INSECTS WITHOUT WINGS

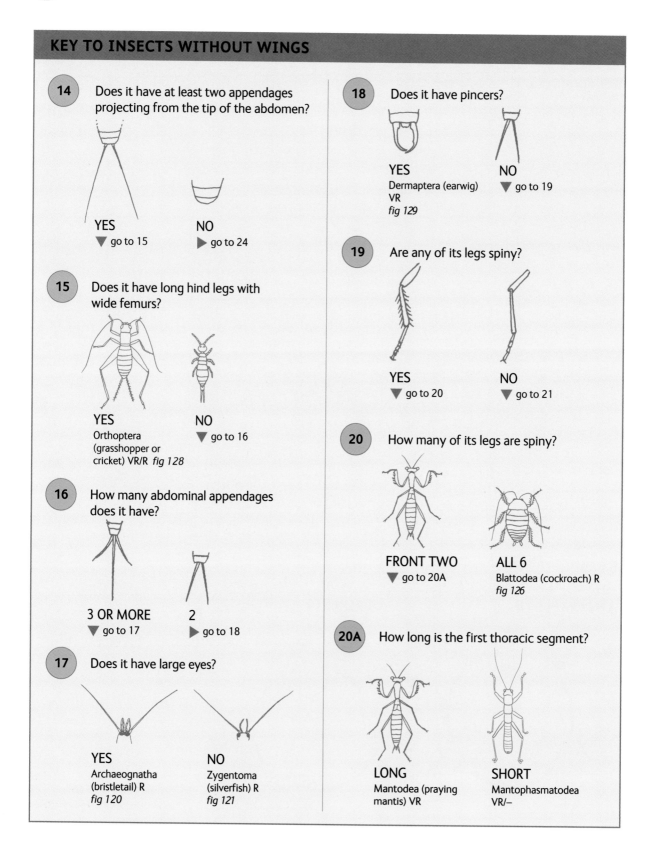

14 Does it have at least two appendages projecting from the tip of the abdomen?

YES
▼ go to 15

NO
▶ go to 24

15 Does it have long hind legs with wide femurs?

YES
Orthoptera
(grasshopper or
cricket) VR/R *fig 128*

NO
▼ go to 16

16 How many abdominal appendages does it have?

3 OR MORE
▼ go to 17

2
▶ go to 18

17 Does it have large eyes?

YES
Archaeognatha
(bristletail) R
fig 120

NO
Zygentoma
(silverfish) R
fig 121

18 Does it have pincers?

YES
Dermaptera (earwig)
VR
fig 129

NO
▼ go to 19

19 Are any of its legs spiny?

YES
▼ go to 20

NO
▼ go to 21

20 How many of its legs are spiny?

FRONT TWO
▼ go to 20A

ALL 6
Blattodea (cockroach) R
fig 126

20A How long is the first thoracic segment?

LONG
Mantodea (praying
mantis) VR

SHORT
Mantophasmatodea
VR/–

KEY TO INSECTS WITHOUT WINGS

21 Are the tips of its forelegs enlarged?

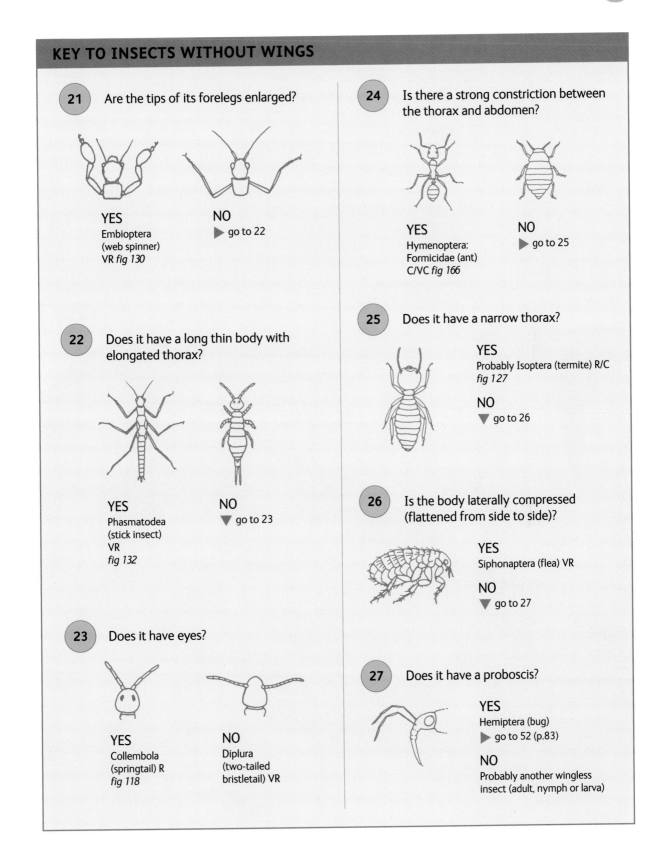

YES
Embioptera
(web spinner)
VR *fig 130*

NO
▶ go to 22

22 Does it have a long thin body with elongated thorax?

YES
Phasmatodea
(stick insect)
VR
fig 132

NO
▼ go to 23

23 Does it have eyes?

YES
Collembola
(springtail) R
fig 118

NO
Diplura
(two-tailed
bristletail) VR

24 Is there a strong constriction between the thorax and abdomen?

YES
Hymenoptera:
Formicidae (ant)
C/VC *fig 166*

NO
▶ go to 25

25 Does it have a narrow thorax?

YES
Probably Isoptera (termite) R/C
fig 127

NO
▼ go to 26

26 Is the body laterally compressed (flattened from side to side)?

YES
Siphonaptera (flea) VR

NO
▼ go to 27

27 Does it have a proboscis?

YES
Hemiptera (bug)
▶ go to 52 (p.83)

NO
Probably another wingless
insect (adult, nymph or larva)

INCOMPLETE METAMORPHOSIS

There are many orders of insects with incomplete metamorphosis in their life-cycle. Some of the orders are extinct, but all of those that occur in amber are extant.

BLATTODEA – COCKROACHES

Cockroaches are dorso-ventrally flattened insects with toughened forewings (tegmina) and a head-shield (pronotum) (fig 125). They have very long antennae, very spiny legs and cerci that are usually hairy. The wings have a dense venation with many parallel and branching veins with few or no cross-veins. The radius sends many parallel branches towards the costa, and the anal veins occur in a distinctive curved or triangular area at the base of the wing known as the clavus. Cockroaches are fairly rare in amber because they are large and strong and can therefore pull themselves free of the resin. Both adults and nymphs occur in amber, but the nymphs do not have wings (fig 126).

BELOW (fig 125) Neopteran: cockroach (Blattodea) in Baltic amber. (Length 12 mm, or ½ in.)

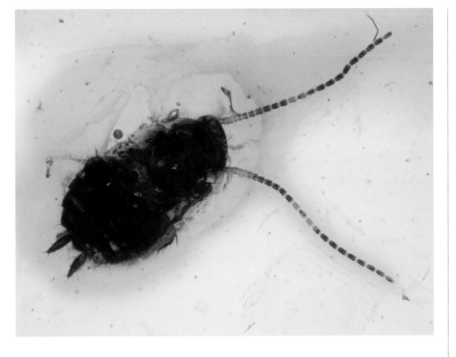

ABOVE (fig 127) Soldier termite (Isoptera) in Dominican amber, which has a nozzle for squirting glue. (Length 3 mm or ⅛ in.)

ISOPTERA – TERMITES

Termites have small eyes and very short cerci that are very difficult to see. They are social insects with winged, worker and soldier castes (see figs 86, 107, 127). The winged castes have membranous wings with a wrinkled or knobbly texture and faint, unpigmented veins. The costa, subcosta and radius are the most distinct veins and run parallel to each other. The workers are wingless and generally have short, fat, segmented bodies. The head is usually wider than the thorax. Termites are common in Dominican amber but rare in Baltic amber and they often have large bubbles projecting from their bodies. These bubbles consist of methane produced by the many bacteria in the gut that helped the termite to digest its food.

BELOW (fig 128) Cricket (Orthoptera: Grylloidea) in Dominican amber. (Length 5.7 mm, or ¼ in.)

ORTHOPTERA – GRASSHOPPERS, CRICKETS AND LOCUSTS

Orthoptera possess hind legs that have wide femurs and are much longer than the other legs. These adaptations enable them to jump. Orthoptera are rare in amber, though crickets (Superfamily Grylloidea) and several kinds of grasshopper have been recorded. Crickets have short, fat bodies with long, hairy cerci and long antennae (fig 128). They sometimes have wings that lie flat over the body. In males the wing veins are convoluted to enable singing, whereas in females the veins are straight and have many cross-veins. By contrast, grasshoppers generally have longer bodies, short cerci and their wings are held vertically along their bodies. Some female Orthoptera have a long, curved, pointed ovipositor projecting from the abdomen, which is used for laying eggs.

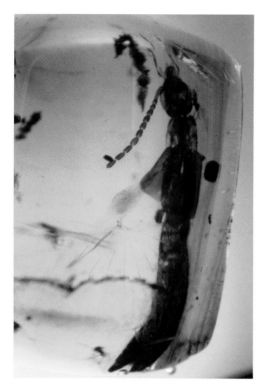

DERMAPTERA – EARWIGS

Earwigs have elongate bodies with cerci that are modified to form pincers (figs 77, 129). They have short wing cases (elytra) with a fan-like hindwing folded underneath. The antennae are generally long. Earwigs are extremely rare in amber, which is surprising considering they are commonly found in cracks in the bark of trees today.

EMBIOPTERA – WEB SPINNERS

Web spinners are small (a few millimetres long) with an elongate body, small asymmetrical cerci and the ends of their front legs are often greatly enlarged. Some have wings, others are wingless. The winged forms have two pairs of equal-sized wings with a few longitudinal veins (fig 130). They are a tropical group that is extremely rare in amber.

MANTODEA – PRAYING MANTISES

Praying mantises have elongate bodies and a wide head with large eyes and long antennae. Their front legs are held upwards and bent by at least 90 degrees. These legs are used to catch prey and often have many spines. Mantids are extremely rare in amber.

PHASMATODEA – STICK OR LEAF INSECTS

Phasmatodeans are either very long and thin (often wingless) to mimic twigs or flat and wide to mimic leaves. They are very rare in amber, and only the long and thin kind have been found (fig 132).

MANTOPHASMATODEA – ROCK CRAWLERS, GLADIATORS OR HEEL WALKERS

This wingless group was only named recently. They are similar to Mantodea in having spiny bent forelegs but the first segment of the thorax is short. They are extremely rare and are only found in Baltic amber.

TOP (fig 129) Earwig (Dermaptera) in Dominican amber. (Length 7 mm, or ¼ in.)

ABOVE (fig 130) Web spinner (Embioptera) in Colombian copal. (Length 6 mm or ¼ in.)

RIGHT (fig 131) Stonefly (Plecoptera: Leuctridae) in Baltic amber. (Length 8 mm, or ⁵⁄₁₆ in.)

OPPOSITE (fig 132) Stick insect (Phasmatodea) in Baltic amber. (Length 11 mm, or ⁷⁄₁₆ in.)

PLECOPTERA – STONEFLIES

Stoneflies have long antennae, long cerci and two pairs of equal-sized wings. The venation is distinctive in that there are many cross-veins between the median and cubital veins, giving a ladder-like appearance (fig 131). They are very rare in amber. Most of those in Baltic amber belong to the family Leuctridae, which wrap their wings around their body when at rest.

PSOCOPTERA – BARKLICE AND BOOKLICE

Psocopterans are very small (a couple of millimetres long) with two pairs of wings held roof-like over the short body and a squarish head that has long thin antennae. The forewings are slightly larger than the hindwings and often have a large pterostigma, or they can be pointed and hairy (fig 133). They are fairly common in Dominican amber but very rare in Baltic amber, and many families have been recorded.

THYSANOPTERA – THRIPS

Thrips are very small (usually less than 2 mm, or ¹⁄₁₆ in, long), thin and the adults have two pairs of wings. The wings are of equal size, narrow with one or two veins and have a distinctive fringe of hairs (fig 134). They are rare in amber but have probably been overlooked because of their small size. The extinct lophioneurids (see fig 46, p.27) are a bizarre family and not typical thrips.

ZORAPTERA

The order Zoraptera are small, slightly hairy insects with an elongate thorax. They are either apterous (wingless) or have two pairs of wings with a few veins. The apterous ones do not have any eyes and are blind. Only a few Zoraptera are known from Dominican and Burmese amber.

PHTHIRAPTERA – LICE

Lice are apterous, flattened, parasitic insects with clawed feet. The only record of this order in amber is of eggs on mammal hair in Baltic amber.

KEY TO WINGED INSECTS I

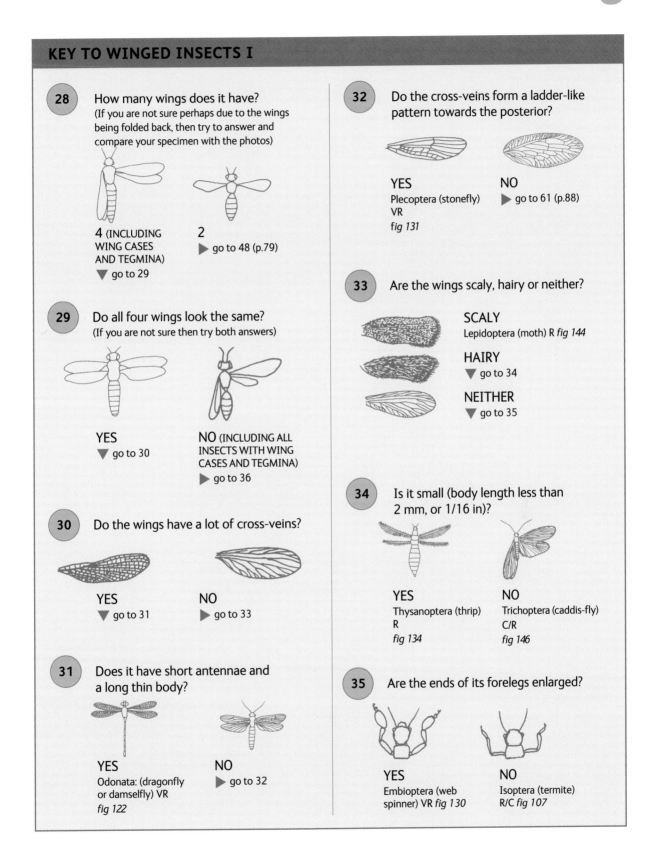

28 How many wings does it have?
(If you are not sure perhaps due to the wings being folded back, then try to answer and compare your specimen with the photos)

4 (INCLUDING WING CASES AND TEGMINA)
▼ go to 29

2
▶ go to 48 (p.79)

29 Do all four wings look the same?
(If you are not sure then try both answers)

YES
▼ go to 30

NO (INCLUDING ALL INSECTS WITH WING CASES AND TEGMINA)
▶ go to 36

30 Do the wings have a lot of cross-veins?

YES
▼ go to 31

NO
▶ go to 33

31 Does it have short antennae and a long thin body?

YES
Odonata: (dragonfly or damselfly) VR
fig 122

NO
▶ go to 32

32 Do the cross-veins form a ladder-like pattern towards the posterior?

YES
Plecoptera (stonefly) VR
fig 131

NO
▶ go to 61 (p.88)

33 Are the wings scaly, hairy or neither?

SCALY
Lepidoptera (moth) R fig 144

HAIRY
▼ go to 34

NEITHER
▼ go to 35

34 Is it small (body length less than 2 mm, or 1/16 in)?

YES
Thysanoptera (thrip) R
fig 134

NO
Trichoptera (caddis-fly) C/R
fig 146

35 Are the ends of its forelegs enlarged?

YES
Embioptera (web spinner) VR fig 130

NO
Isoptera (termite) R/C fig 107

KEY TO WINGED INSECTS I

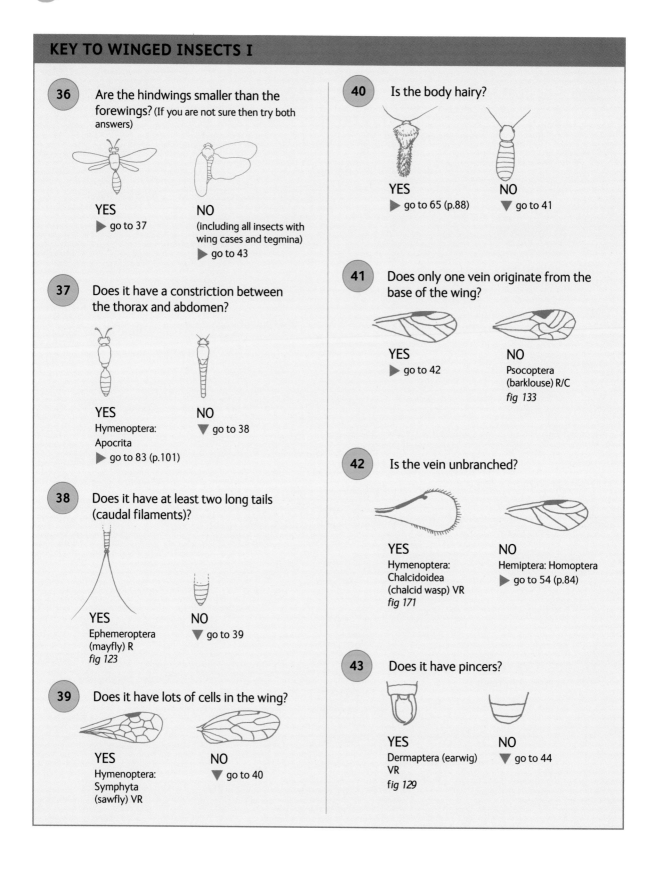

36 Are the hindwings smaller than the forewings? (If you are not sure then try both answers)

YES
▶ go to 37

NO
(including all insects with wing cases and tegmina)
▶ go to 43

37 Does it have a constriction between the thorax and abdomen?

YES
Hymenoptera: Apocrita
▶ go to 83 (p.101)

NO
▼ go to 38

38 Does it have at least two long tails (caudal filaments)?

YES
Ephemeroptera (mayfly) R
fig 123

NO
▼ go to 39

39 Does it have lots of cells in the wing?

YES
Hymenoptera: Symphyta (sawfly) VR

NO
▼ go to 40

40 Is the body hairy?

YES
▶ go to 65 (p.88)

NO
▼ go to 41

41 Does only one vein originate from the base of the wing?

YES
▶ go to 42

NO
Psocoptera (barklouse) R/C
fig 133

42 Is the vein unbranched?

YES
Hymenoptera: Chalcidoidea (chalcid wasp) VR
fig 171

NO
Hemiptera: Homoptera
▶ go to 54 (p.84)

43 Does it have pincers?

YES
Dermaptera (earwig) VR
fig 129

NO
▼ go to 44

KEY TO WINGED INSECTS I

44 Does it have a proboscis?

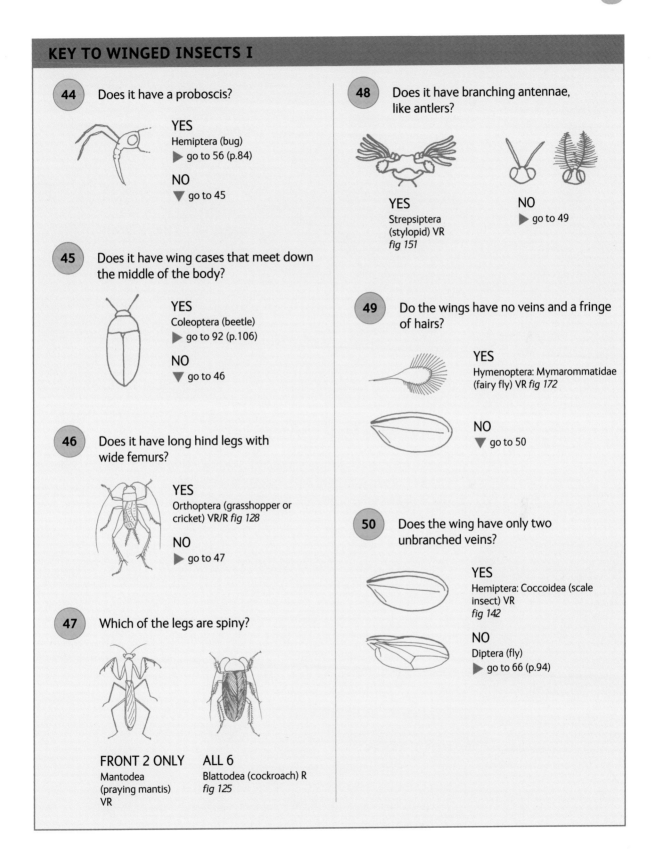

YES
Hemiptera (bug)
▶ go to 56 (p.84)

NO
▼ go to 45

45 Does it have wing cases that meet down the middle of the body?

YES
Coleoptera (beetle)
▶ go to 92 (p.106)

NO
▼ go to 46

46 Does it have long hind legs with wide femurs?

YES
Orthoptera (grasshopper or cricket) VR/R *fig 128*

NO
▶ go to 47

47 Which of the legs are spiny?

FRONT 2 ONLY
Mantodea
(praying mantis)
VR

ALL 6
Blattodea (cockroach) R
fig 125

48 Does it have branching antennae, like antlers?

YES
Strepsiptera
(stylopid) VR
fig 151

NO
▶ go to 49

49 Do the wings have no veins and a fringe of hairs?

YES
Hymenoptera: Mymarommatidae
(fairy fly) VR *fig 172*

NO
▼ go to 50

50 Does the wing have only two unbranched veins?

YES
Hemiptera: Coccoidea (scale insect) VR
fig 142

NO
Diptera (fly)
▶ go to 66 (p.94)

BUGS (HEMIPTERA)

Insects in the order Hemiptera comprise the bugs. They have a tube-like proboscis that they use to suck plant sap or blood. There are two suborders: the Heteroptera and Homoptera, although some scientists use the word bug for the Heteroptera only.

HETEROPTERA

The Heteroptera have a curved proboscis that projects downwards from the front of the head and long antennae consisting of only a few segments. They also have forewings that are usually divided into two parts. The basal half of the wing is thickened (sclerotized) whereas the distal half is membranous. The membranous parts of the forewings overlap when the wings are folded (fig 135). Heteroptera are fairly rare in amber although many families have been recorded. The families generally appear similar to each other and are difficult to identify. Members of a distinctive subfamily of assassin bugs (Reduviidae: Emesinae) have an elongate abdomen and very long legs (fig 136).

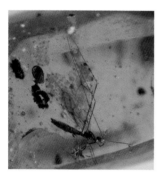

ABOVE (fig 136) Emesine assassin bug (Heteroptera: Reduviidae: Emesinae) in Dominican amber. (Length of forewing 3 mm or 1/8 in.)

HOMOPTERA

Members of the suborder Homoptera have a straight proboscis that originates at the back of the head and projects under the body. The forewings are not separated into two parts. Homoptera are fairly common in amber and the different superfamilies are fairly easy to distinguish.

LEFT (fig 137) Leafhopper
(Homoptera: Cicadellidae) in
Dominican amber. (Length 3 mm,
or ⅛ in.)

The Fulgoroidea, Cicadelloidea and Cercopoidea are similar in that their forewings are sclerotized and held roof-like over the body. They have extremely short antennae that are barely visible. The forewings have a distinctive triangular clavus and are often patterned. The Fulgoroidea (commonly known as planthoppers) are large bugs with three distinctive ridges running down the front of the head, a few spines on the legs and the two anal veins in the clavus merged to form a Y-shape (figs 70, 138). They are fairly rare in amber.

BELOW (fig 138) Planthopper
(Homoptera: Fulgoroidea) in Baltic
amber. (Length of forewing 6.5 mm,
or ¼ in.)

The Cicadelloidea (commonly known as leaf or treehoppers) have a triangular head and two anal veins in the clavus that do not merge. The family Cicadellidae (commonly known as leafhoppers) have spiny hind legs and are fairly common in Dominican amber but rare in Baltic amber (figs 84, 137). The family Membracidae (commonly known as treehoppers) have large head-shields of various shapes and sizes, and are known from Dominican amber, where they are extremely rare. The superfamily Cercopoidea (commonly known as froghoppers or spittlebugs) do not have any veins visible in the clavus and are extremely rare in amber (fig 139).

The superfamilies Aphidoidea, Psylloidea, Aleyrodoidea and Coccoidea are small insects (generally less than 3 mm, or ⅛ in, long) with membranous wings that only have a few veins. The Aphidoidea are known commonly as aphids or greenfly. They have fat bodies and sometimes hold their wings vertically when at rest. The forewings are much larger than the hindwings, have a pterostigma and a prominent vein running close to the anterior edge of the wing with a few veins coming obliquely off it (fig 140). They often have tubes on the body, which would have ejected blood to repel predators (fig 141) and they frequently have an extremely long proboscis that can be twice the length of the body. They are fairly common in Baltic amber but extremely rare in Dominican amber.

The Psylloidea (commonly known as jumping plant lice) have forewings that are slightly larger than the hindwings, and possess short antennae. The forewings have a single main vein that branches a few times. They are extremely rare in amber. The Aleyrodoidea are known commonly as whiteflies because their wings are white. They have two pairs of wings with a single vein that forks near the middle of the wing (fig 143). They are extremely rare in amber.

The Coccoidea are known commonly as scale insects. Only the males have wings, of which only a single pair is present (see fig 47, p.27). The wings usually have two unbranched veins: the first vein runs close to the anterior margin, the second runs through the middle of the wing (fig 142). The wing often has a wrinkled texture. Scale insects are very rare in amber.

ABOVE (fig 139) Froghopper (Homoptera: Cercopoidea) in Baltic amber. (Length 6.5 mm, or ¼ in.)

BELOW (fig 140) Winged aphid (Homoptera: Aphidoidea) in Baltic amber. (Length 3.5 mm, or ⅛ in.)

BELOW RIGHT (fig 141) Aphid (Homoptera: Aphidoidea) in Baltic amber. (Length 1.7 mm, or 1/16 in.)

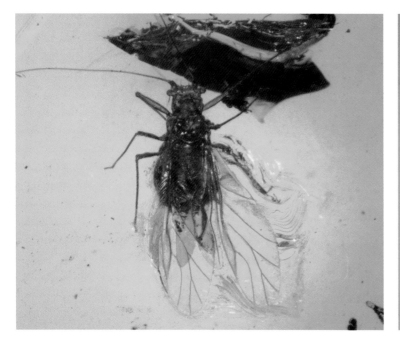

LEFT (fig 142) Scale insect (Homoptera: Coccoidea) in Baltic amber. (Length of wing 1.2 mm, or $\frac{1}{16}$ in.)

BELOW (fig 143) Whitefly (Homoptera: Aleyrodoidea) in Burmese amber. (Length 1 mm or $\frac{1}{32}$ in.)

KEY TO BUGS

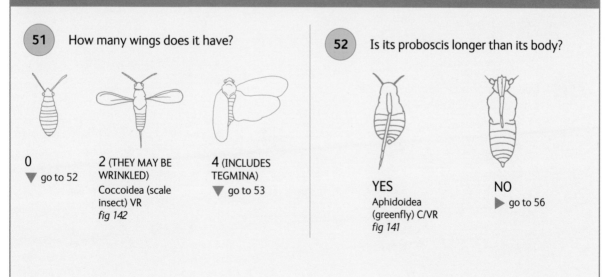

51   How many wings does it have?

0
▼ go to 52

2 (THEY MAY BE WRINKLED)
Coccoidea (scale insect) VR
fig 142

4 (INCLUDES TEGMINA)
▼ go to 53

52   Is its proboscis longer than its body?

YES
Aphidoidea (greenfly) C/VR
fig 141

NO
▶ go to 56

KEY TO BUGS

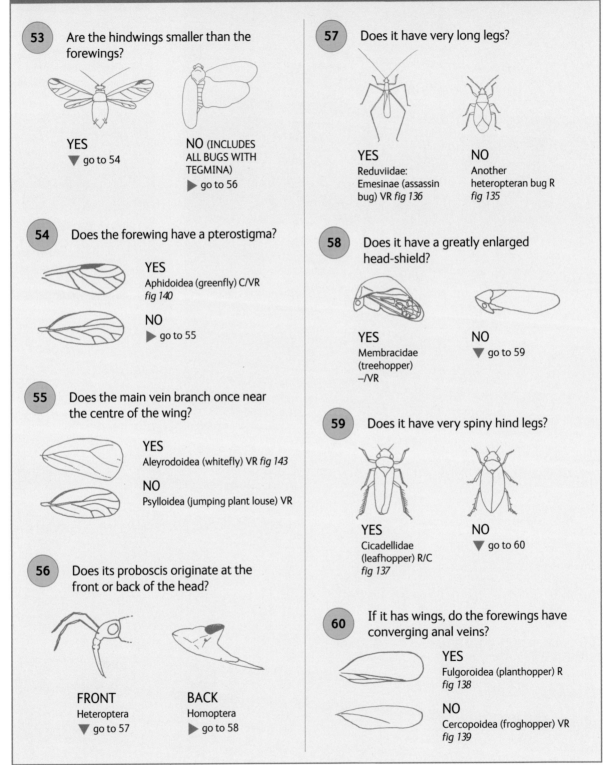

53 Are the hindwings smaller than the forewings?

YES
▼ go to 54

NO (INCLUDES ALL BUGS WITH TEGMINA)
▶ go to 56

54 Does the forewing have a pterostigma?

YES
Aphidoidea (greenfly) C/VR *fig 140*

NO
▶ go to 55

55 Does the main vein branch once near the centre of the wing?

YES
Aleyrodoidea (whitefly) VR *fig 143*

NO
Psylloidea (jumping plant louse) VR

56 Does its proboscis originate at the front or back of the head?

FRONT
Heteroptera
▼ go to 57

BACK
Homoptera
▶ go to 58

57 Does it have very long legs?

YES
Reduviidae: Emesinae (assassin bug) VR *fig 136*

NO
Another heteropteran bug R *fig 135*

58 Does it have a greatly enlarged head-shield?

YES
Membracidae (treehopper) –/VR

NO
▼ go to 59

59 Does it have very spiny hind legs?

YES
Cicadellidae (leafhopper) R/C *fig 137*

NO
▼ go to 60

60 If it has wings, do the forewings have converging anal veins?

YES
Fulgoroidea (planthopper) R *fig 138*

NO
Cercopoidea (froghopper) VR *fig 139*

COMPLETE METAMORPHOSIS

The endopterygotes are those insects with complete metamorphosis. There are eleven extant orders, all of which have been found in amber.

LEPIDOPTERA – MOTHS AND BUTTERFLIES

Moths and butterflies usually have two pairs of scaly wings, a long, curled proboscis and a hairy body. Moths hold their wings back along their body whereas butterflies are generally larger and hold their wings vertically above the body when at rest. Moths are fairly rare in amber and butterflies are extremely rare, although the latter are sometimes seen as fakes. The scales on the wings refract the light to produce the bright colour patterns. However, the amber penetrates the wings and original colour patterns are lost or very faded (fig 144). (Some other insects, such as planthopper bugs (see fig 70, p.38), do have colour patterns preserved but this is because the wings are pigmented, which is not affected by the amber.) It is common to find many of the scales embedded in the surrounding amber, which would have detached while the moth was struggling to escape. Sometimes the proboscis is partially uncurled in the amber. Many different families of moths have been recorded in amber, but they are often difficult to identify. Caterpillars also occur in amber (fig 145).

BELOW LEFT (fig 144) Moth (Lepidoptera) in Baltic amber. (Length 5.5 mm, or ³⁄₁₆ in.)

BELOW (fig 145) Caterpillar (Lepidoptera) in Baltic amber. (Length 8.5 mm, or ⁵⁄₁₆ in.)

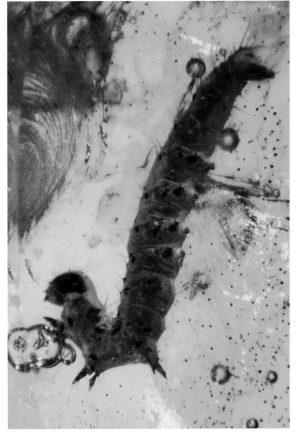

TRICHOPTERA — CADDIS-FLIES

Caddis-flies have two pairs of wings, which they hold roof-like over the body. Caddis-flies are very similar in appearance to moths except they have hairy wings rather than scaly, and they have chewing mouthparts rather than a proboscis (fig 146). They are common in Baltic amber but fairly rare in Dominican amber. Many different families have been recorded but they are difficult to tell apart.

RIGHT (fig 146) Caddis-fly (Trichoptera) in Baltic amber. (Length 8 mm, or ⁵/₁₆ in.)

RIGHT (fig 147) Scorpion fly (Mecoptera) in Baltic amber. (Length of wing 11 mm, or ⁷/₁₆ in.)

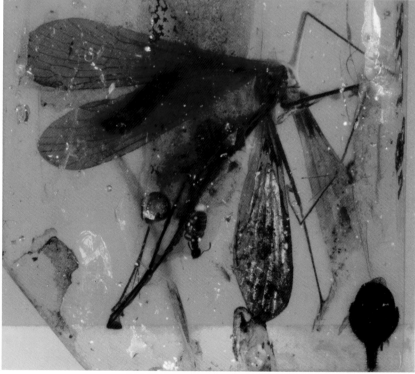

LEFT (fig 148) Lacewing (Neuroptera) in Baltic amber. (Length 8 mm, or 5/16 in.)

BELOW (fig 149) Mantis fly (Neuroptera: Mantispidae) in English (Baltic) amber. (Length 6 mm or ¼ in.)

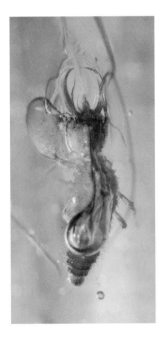

MECOPTERA – SCORPION FLIES

Scorpion flies have two pairs of equal-sized wings with veins that progressively branch and have cross-veins between them (fig 147). Scorpion flies have a characteristic elongate head that points downwards, long antennae and the males often have a long tail that curves upwards. They are extremely rare in Baltic amber only and three families have been recorded.

NEUROPTERA – LACEWINGS AND ANT-LIONS

Lacewings and ant-lions have two pairs of equal-sized wings with many branching veins and cross-veins. Most of the veins branch immediately before they reach the margin (fig 148). They are very rare in amber but several families have been recorded. Larvae with large jaws are also known (fig 150). One distinctive family (Mantispidae), known as mantis flies, have long, bent spiny forelegs similar to those of a praying mantis (fig 149). It is surprising that Neuroptera are rare in Baltic amber because some families feed on aphids, which are common in this amber.

MEGALOPTERA – ALDER FLIES

Alder flies appear similar to lacewings except they have much thicker wing veins. They are extremely rare in Baltic amber only, and three families have been recorded. The specimen in fig 69, p.38 is the only known example of the extinct family Corydasialidae, which was named recently.

ABOVE (fig 150) Lacewing larva (Neuroptera) in Baltic amber. (Length including jaws 3 mm, or ⅛ in.)

ABOVE (fig 151) Stylopid (Strepsiptera) in Dominican amber. (Length 1.5 mm, or ¹⁄₁₆ in.)

RAPHIDIOPTERA – SNAKE FLIES

Snake flies also appear similar to lacewings except they have an elongated thorax. Like alder flies they are extremely rare in Baltic amber and have not been recorded in Dominican amber.

SIPHONAPTERA – FLEAS

Fleas are specialized vertebrate parasites. They are very small (1 mm, or ¹⁄₃₂ in, or less), wingless and have a laterally compressed body, which is hairy. They are extremely rare in amber.

STREPSIPTERA – STYLOPIDS

Stylopids are very small (typically 1 mm, or ¹⁄₃₂ in, long) with a single pair of fan-like wings and branching antennae that look like antlers (fig 151). They are specialized parasites of other insects and are extremely rare in amber.

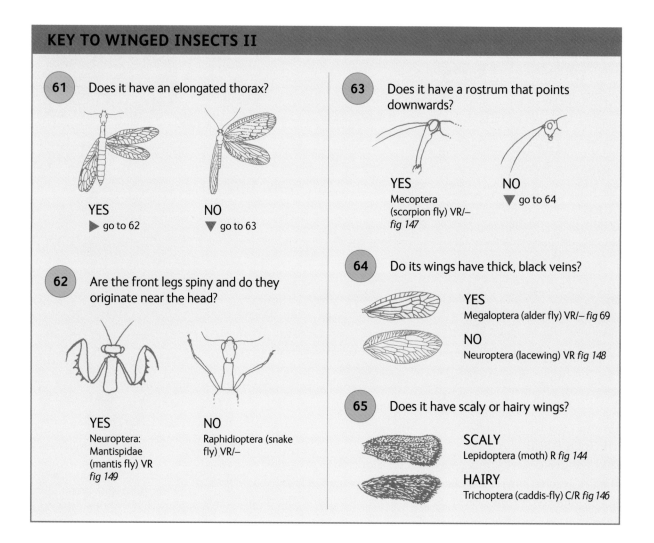

KEY TO WINGED INSECTS II

61 Does it have an elongated thorax?

YES
▶ go to 62

NO
▼ go to 63

62 Are the front legs spiny and do they originate near the head?

YES
Neuroptera: Mantispidae (mantis fly) VR fig 149

NO
Raphidioptera (snake fly) VR/–

63 Does it have a rostrum that points downwards?

YES
Mecoptera (scorpion fly) VR/– fig 147

NO
▼ go to 64

64 Do its wings have thick, black veins?

YES
Megaloptera (alder fly) VR/– fig 69

NO
Neuroptera (lacewing) VR fig 148

65 Does it have scaly or hairy wings?

SCALY
Lepidoptera (moth) R fig 144

HAIRY
Trichoptera (caddis-fly) C/R fig 146

TRUE FLIES (DIPTERA)

The order Diptera of the class Insecta are the flies, which are characterized by only having a single pair of wings. The hindwings are reduced to small club-like structures known as halteres, which are used as balancers during flight. Flies are the most abundant insects in Baltic amber and even swarms are preserved (see fig 2, p.5).

There are three suborders of Diptera. The Nematocera are easy to identify in that they have long, multi-segmented antennae. The Brachycera and Cyclorrhapha (=Muscomorpha) have short antennae, often with a bristle known as an arista.

NEMATOCERA

The Nematocera are the most primitive suborder of flies and are the most common insects in Baltic amber. The males have external genitalia consisting of a pair of hooks. Many families have been recorded and the groups most likely to be encountered are briefly described here.

The superfamily Tipuloidea are commonly known as crane-flies or daddy-long-legs (the latter is also used for harvestmen spiders). They are easy to identify in that they are generally large flies with extremely long legs, a long slender body, an elongate head and slender wings (see fig 71, p.38). Often, legs have broken away from the body due to the insect struggling. The wings have many veins and frequently have a squarish cell towards the tip of the wing. They are fairly rare in amber.

The superfamily Mycetophiloidea are known commonly as fungus gnats. They generally have long antennae and rounded wings (fig 152). Their bodies are often long and slender, but they can be short and fat. This group is very diverse and many different families and genera have been recorded. The wing venation is distinct and both the median and cubital veins branch once, usually in the centre of the wing. The radius and radial sector are connected by one or two cross-veins. The radial sector terminates at the tip of the wing and sometimes sends a branch to the anterior margin, near the tip. They are fairly common in amber. The Sciaridae are a distinctive family of fungus gnats. They can be identified in that only the median vein branches in the centre of the wing and those branches are strongly curved (fig 153). The cubitus branches near the base of the wing. There is only one cross-vein between the radius and radial sector, and the radial sector does not branch. Sciarids are very common in Baltic amber but rare in Dominican amber.

The Scatopsidae are the scavenger flies. They have short bodies, fairly short antennae and rounded wings (fig 72, p.38). The venation is similar to that of sciarids in that only the median branches in the centre of the wing and those branches are

TOP (fig 152) Fungus gnat (Nematocera: Mycetophiloidea) in Baltic amber. This specimen is a male, as shown by its external genitalia. (Length 5.5 mm, or 3/16 in.)

ABOVE (fig 153) Fungus gnat (Nematocera: Sciaridae) in Baltic amber. This specimen is a female, as shown by its tapered abdomen. (Length 3 mm, or 1/8 in.)

ABOVE (fig 154) Pair of mating midges (Nematocera: Chironomidae) in Baltic amber with two air bubbles. (Length of male on right with feathery antennae 1.6 mm, or ¹/₁₆ in.)

ABOVE RIGHT (fig 155) Gall midge (Nematocera: Cecidomyiidae) in Baltic amber. (Length 1.8 mm, or ¹/₁₆ in.)

curved. The venation differs from sciarids in that the radius and radial sector curve upwards and terminate near the centre of the anterior margin of the wing. They are fairly common in Dominican amber but very rare in Baltic amber.

The Chironomidae, Ceratopogonidae and Cecidomyiidae are the midges, which are generally only a couple of millimetres long. The wing venation is useful in telling them apart, but because of their small size this isn't easy to see without a microscope.

The chironomids are the non-biting midges, which have slender bodies and slender wings (fig 154). The males have extremely feathery antennae. The radius is straight, running parallel to the anterior margin, and is connected to the radial sector by a single cross-vein. The radial sector is straight and terminates near the tip of the wing. The median and cubitus veins are faint. The cubitus forks once but the median does not branch. Chironomids are very common in Baltic amber, but rare in Dominican amber.

The ceratopogonids are the biting midges. They differ from chironomids in that the body is shorter and fatter and the wings are rounded (see fig 103, p.53). They also have elongate mouthparts projecting downwards. The radius and radial sector are connected by two cross-veins and they curve upwards and terminate near the centre of the anterior margin. Both the median and cubitus veins fork once. Ceratopogonids are fairly common in amber.

The cecidomyiids are the gall midges. They are similar to ceratopogonids in that they have fat bodies and rounded wings; however, they have much longer antennae with bead-like segments (fig 155). The wing venation is similar to that of chironomids in that the radial sector terminates at the tip of the wing. The median is often absent, and if present it does not usually branch. Usually cecidomyiid wings are hairy; however, some species of chironomids and ceratopogonids also have hairy wings. Cecidomyiids are rare in Baltic amber, but fairly common in Dominican amber.

The Psychodidae are commonly known as moth flies, owl-midges or sandflies. They are small, hairy flies with bead-like antennal segments (fig 156). The wings are generally hairy with many equally spaced, parallel veins. The subfamily Phlebotominae

are the bloodsucking sandflies, which have elongate mouthparts (see fig 104, p.54). Psychodids are rare in Baltic amber, but fairly common in Dominican amber.

Apart from the biting midges and sandflies, there are two other bloodsucking nematoceran families. They are the Culicidae (mosquitoes) and Simuliidae (black-flies). Mosquitoes generally appear similar to chironomid midges in that they have long slender bodies and slender wings (see fig 101, p.52). They differ from chironomids in that mosquitoes are larger, with a long, thin proboscis and scaly wings. The wing venation differs in that the radial sector and median veins branch. Many people think that mosquitoes are common in amber, but this is not the case. Only a few are known in Baltic amber and a few tens of specimens are known in Dominican amber. Most so-called 'mosquitoes' in amber actually belong to other groups; most turn out to be fungus gnats. The black-flies – not to be confused with blackfly (Hemiptera: Aphidoidea) – have short, fat bodies and short antennae (see fig 105, p.54). The wings appear similar to those of mycetophilids in that they are rounded. They also have a radial sector that terminates at the wing tip and a median vein that forks. Simuliids differ from mycetophilids in that the median forks very close to the point where the radial sector is connected to the radius (by a cross-vein) rather than in the centre of the wing. The cubital veins originate right at the base of the wing and one of them is often kinked. Simuliids are very rare in Baltic amber.

BRACHYCERA

The suborders Brachycera and Cyclorrhapha are flies that usually have very short antennae. The Brachycera generally have slender antennae consisting of a few segments with the hair-like arista originating from the tip.

Many families of Brachycera have been recorded in amber but only three are regularly encountered. The Rhagionidae are commonly known as snipe flies. They are large, robust flies with rounded wings that usually possess a pterostigma (fig 157). The wing has many veins and a distinctive medial cell with three or four veins

ABOVE LEFT (fig 156) Moth fly in Baltic amber (Nematocera: Psychodidae). (Length 1.3 mm, or 1/16 in.)

ABOVE (fig 157) Snipe flies (Brachycera: Rhagionidae) in Baltic amber. (Length of right fly 5 mm, or 3/16 in.)

projecting from it that terminate on the posterior margin. Rhagionids are fairly rare in Baltic amber and extremely rare in Dominican amber.

There are several families that have been recorded from amber that are closely related to rhagionids, but they are extremely rare. Of these the Rachiceridae differ from rhagionids in having long multi-segmented antennae (see fig 85, p.44). The Tabanidae are the bloodsucking horseflies. They are larger than rhagionids and have a widely diverging forked radial sector at the tip of the wing (see fig 102, p.52).

The Empididae are commonly known as dance flies. They are generally smaller than rhagionids and often have a long pointed proboscis projecting down from the head (fig 158). The wing venation is diverse in this group. The radial sector terminates at the tip of the wing and is either unbranched or forks near the tip. There are several subfamilies that have an elongate medial cell with two or three veins projecting from it. Most also have a cubital vein that does not reach the posterior margin; instead two cross-veins project horizontally from it to produce an oblique, upside-down T-shape. Some species have neither the medial cell or T-shaped cubitus and these belong to the subfamily Tachydromiinae. They have a simpler venation with an unbranched median that produces two cross-veins. One cross-vein connects to the radial sector and the other to the cubitus. The cross-veins either originate at the same point near the centre of the wing or are slightly staggered. Empidids are fairly rare in amber.

The Dolichopodidae are small flies that are known as long-legged flies, although their legs are not noticeably longer than those of other groups. In the males the external genitalia are large and are often curved under the body. The wings have a simple venation with a few unbranched veins (fig 159). There is a single, distinctive cross-vein between the median and cubitus situated towards the posterior margin. Sometimes the median is kinked before it terminates at the tip of the wing. Dolichopodids are very common in Baltic amber but fairly rare in Dominican amber.

CYCLORRHAPHA

The Cyclorrhapha are the most advanced suborder of flies and are the most common group living today; however, they are rare in amber. They generally have blob-like antennae with the arista originating near the base. Sometimes the arista of cyclorrhaphans is hairy. Many families have been recorded but they are very difficult to tell apart by a non-expert. Most have a simple venation very similar to that of dolichopodids except they also have a cross-vein between the median and radial sector near the centre of the wing (fig 160). One such family is the Drosophilidae (fruit flies), which is only known in the fossil record from amber (see fig 83, p.43). A recognizable family is the Phoridae, which are commonly known as scuttle flies. They are small, black, hairy flies that have spiny palps (sensory appendages next to the mouth) and rounded

ABOVE (fig 160) Advanced fly (Cyclorrapha) in Baltic amber. (Length 4.3 mm, or ³⁄₁₆ in.)

wings with a distinctive venation (figs 106, 161). The radius and radial sector are thick and terminate on the anterior margin. The four other veins originate from the radial sector or the base of the wing. They are unbranched, equally spaced and curve slightly to terminate at the tip of the wing or on the posterior margin. Phorids are fairly rare in amber. Another distinctive family are the Syrphidae (hoverflies). These are large flies that are generally not hairy. The median and cubitus veins do not reach the posterior margin; instead, they bend upwards and join the vein above (fig 162). They are rare in Baltic amber and extremely rare in Dominican amber.

ABOVE (fig 161) Scuttle fly (Cyclorrapha: Phoridae) in Dominican amber. (Length 1.7 mm, or ¹⁄₁₆ in.)

LEFT (fig 162) Hoverfly (Cyclorrapha: Syrphidae) in Baltic amber. (Length 6 mm, or ¼ in.)

KEY TO TRUE FLIES

66 Are the antennae much longer than the length of the head?

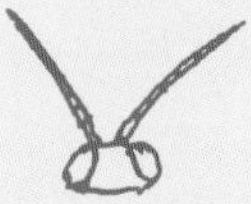

YES
▼ go to 67

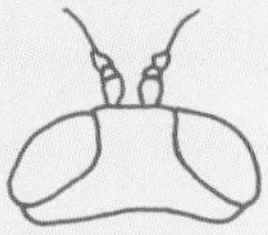

NO
▶ go to 75

67 Does it have slender or rounded wings?

SLENDER
▼ go to 68

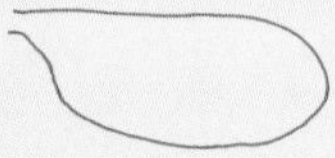

ROUNDED
▼ go to 70

68 Does it have a cell in the wing?

YES
Probably Tipuloidea (crane-fly) R
fig. 62

NO
▼ go to 69

69 Does it have a long proboscis and scaly wings?

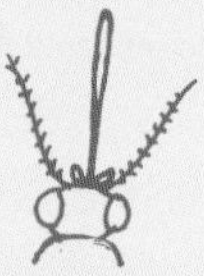

YES
Culicidae (mosquito)
VR/R fig. 90

NO
Probably
Chironomidae
(midge) VC/R fig. 138

70 Does the wing have many equally spaced, parallel veins?

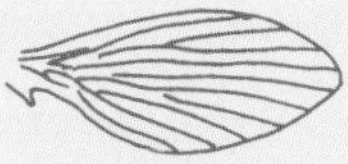

YES
Psychodidae (moth fly) R/C
fig. 140

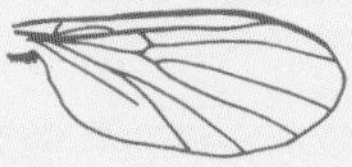

NO
▶ go to 71

71 Does it have long, beaded antennae?

YES
Cecidomyiidae
(gall midge) R/C
fig. 139

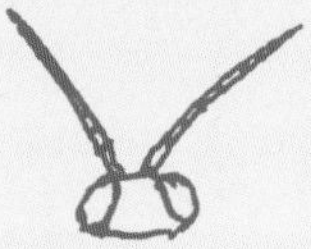

NO
▼ go to 72

72 Does the radial sector reach the wing tip?

YES
▼ go to 73

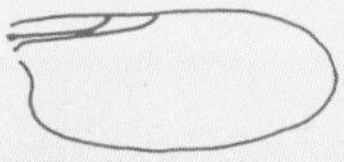

NO
▼ go to 74

73 Do the cubitus or radial sector branch?

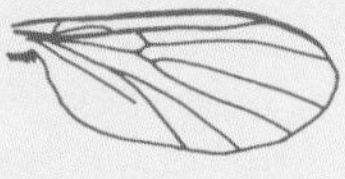

YES
Probably Mycetophiloidea
(fungus gnat) C
fig. 136

NO
Probably Sciaridae (fungus gnat)
VC/R
fig. 137

KEY TO TRUE FLIES

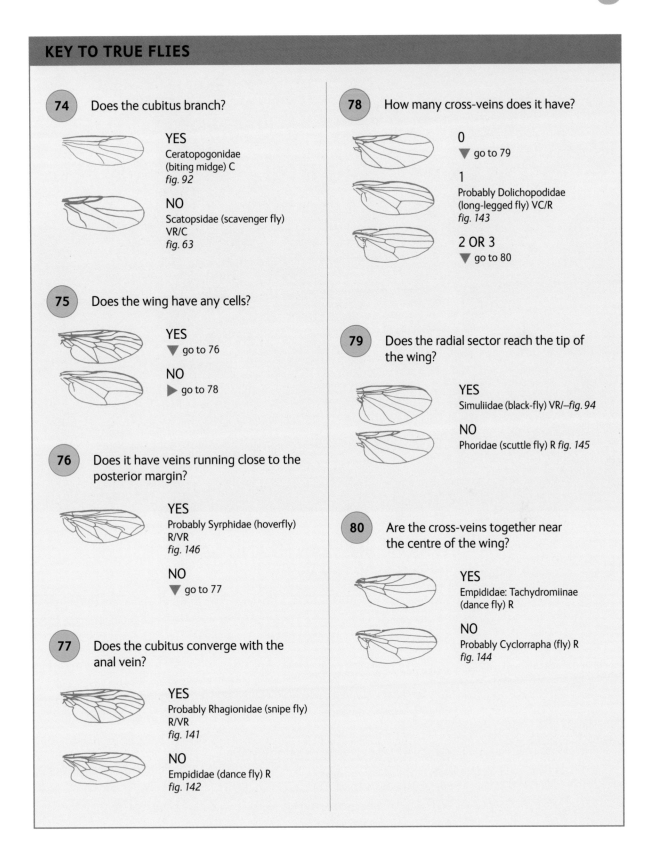

74 Does the cubitus branch?

YES
Ceratopogonidae
(biting midge) C
fig. 92

NO
Scatopsidae (scavenger fly)
VR/C
fig. 63

75 Does the wing have any cells?

YES
▼ go to 76

NO
▶ go to 78

76 Does it have veins running close to the posterior margin?

YES
Probably Syrphidae (hoverfly)
R/VR
fig. 146

NO
▼ go to 77

77 Does the cubitus converge with the anal vein?

YES
Probably Rhagionidae (snipe fly)
R/VR
fig. 141

NO
Empididae (dance fly) R
fig. 142

78 How many cross-veins does it have?

0
▼ go to 79

1
Probably Dolichopodidae
(long-legged fly) VC/R
fig. 143

2 OR 3
▼ go to 80

79 Does the radial sector reach the tip of the wing?

YES
Simuliidae (black-fly) VR/–*fig. 94*

NO
Phoridae (scuttle fly) R *fig. 145*

80 Are the cross-veins together near the centre of the wing?

YES
Empididae: Tachydromiinae
(dance fly) R

NO
Probably Cyclorrapha (fly) R
fig. 144

ABOVE (fig 163) Orchid bee (Aculeata: Apidae: Euglossinae) in Colombian copal. (Length 10 mm, or ⅜ in.)

WASPS, BEES, ANTS AND SAWFLIES (HYMENOPTERA)

Insects in the order Hymenoptera have two pairs of wings with a reduced venation. The forewings usually have a pterostigma and are larger than the hindwings. There are two suborders: the Symphyta and the Apocrita.

SYMPHYTA – SAWFLIES

The Symphyta are large and have many distinct cells in their wings. Sawflies are extremely rare in amber.

APOCRITA – WASPS, BEES AND ANTS

The Apocrita are the wasps, bees and ants. They differ from the Symphyta in that there is usually a constriction between the thorax and abdomen (see fig 45, p.26), and many of them possess stings. Many families have been recorded in amber. The Apocrita are divided into the Aculeata and the Parasitica.

LEFT (fig 164) Digger wasp (Aculeata: Sphecidae) in Baltic amber. (Length 3.8 mm, or ⅛ in.)

ACULEATA

The aculeates are generally large, have four straight veins originating from the base of the wing and several cells in the centre of the wing. The superfamily Apoidea comprise the bees. Bees are easy to identify in that they are fat and usually hairy. The front of the head is flat with two vertically elongate eyes and antennae that bend at 90 degrees in the middle. Bees are very rare in Baltic amber, but there is a species that is common in Dominican amber – *Proplebeia dominicana* (see fig 100, p.51). This is a small stingless bee (4 mm, or ³⁄₁₆ in, long) belonging to the family Apidae that collected tree resin to make its nest and often got trapped while doing this. Sometimes you can see the pollen sacs on its legs, which are full of resin. Bees are important pollinators of flowering plants. A fossil orchid bee (Subfamily Euglossinae) was discovered in Colombian copal that is a striking metallic green colour (fig 163). A closely related family to the bees is the Sphecidae. Sphecids are commonly known as digger wasps and like bees they too are pollinators of flowers. These wasps are more slender than bees and not hairy (fig 51, p.29, fig 164). The forewings usually have two small, almost square cells towards the tip of the wing. Sphecids are very rare in amber.

The Vespidae are the social wasps. Today we recognize them as the large yellow-and-black striped insects that like jam and can sting! Their wings can fold

RIGHT (fig 165) Social wasp
(Aculeata: Vespidae) in Dominican
amber. (Length 13 mm, or ½ in.)

BELOW (fig 166) Ant (Aculeata:
Formicidae) in Baltic amber. (Length
4.1 mm, or ³⁄₁₆ in.)

BELOW RIGHT (fig 167) Flying ant
(Aculeata: Formicidae) in Baltic
amber. (Length 6.5 mm, or ¼ in.)

longitudinally in half when at rest, which prevents them from damaging their wings while in the nest (fig 165). They are extremely rare in amber.

The Formicidae comprise the ants, which are the most abundant insects in Dominican amber. They are also common in Baltic amber. This is not surprising as ants frequently run up and down tree trunks. Ants differ from other Hymenoptera in that they have two constrictions between the thorax and abdomen. They have squarish heads with small round eyes (see fig 49, p.28), large jaws and antennae that bend at 90 degrees in the middle. They, like some bees, are social insects with different castes. The workers are easy to identify in that they do not have wings (fig 166). Winged ants (kings and queens) have a distinctive venation. Most have a cross-shape formed by two veins intersecting towards the tip of the wing and many also have a square cell in the centre of the wing, orientated in a diamond shape (fig 167).

PARASITICA

The Parasitica are generally much smaller than aculeates. Most are only 2 mm (¹⁄₁₆ in) or less in length (see fig 59, p.33) and usually have a single unbranched vein along the anterior edge of the wing. Many different families have been recorded in amber but they require a powerful microscope to examine them and they are very difficult to tell apart by a non-expert.

The largest and most distinctive of the parasitic wasps belong to the superfamily Ichneumonoidea. These have slender bodies, long antennae and often have a long pointed ovipositor projecting from the tip of the abdomen. Those with long ovipositors parasitize larvae living in wood. The wing venation is less reduced than in other parasitic wasps with three straight veins originating from the base of the forewing. The superfamily consists of two families. The Ichneumonidae (ichneumon wasps) generally have two cells below the pterostigma (fig 168). One of the cells

BELOW LEFT (fig 168) Ichneumon wasp (Parasitica: Ichneumonidae) in Baltic amber. (Length 6 mm, or ¼ in.)

BELOW (fig 169) Braconid wasp (Parasitica: Braconidae) in Baltic amber. (Length excluding ovipositor 3.4 mm, or ⅛ in.)

<small>ABOVE</small> (fig 170) Fairy fly (Parasitica: Mymaridae) in Dominican amber. (Length 0.25 mm, or 1/100 in.)

<small>BELOW</small> (fig 171) Chalcid wasp (Parasitica: Chalcidoidea) in Baltic amber. (Length 1 mm, or 1/32 in.)

is large and curved and may have an additional tiny cell at its tip. The Braconidae (braconid wasps) by contrast have three cells below the pterostigma (fig 169). Ichneumonoids are rare in amber.

The Mymaridae and the Mymarommatidae are commonly known as fairy flies because they have a rim of long hairs around the edge of the wing. Mymarids have four elongate wings whereas mymarommatids have only two rounded wings. Both are parasites of the eggs of other insects. The mymarids are the smallest known insects; the smallest living today is 0.14 mm (1/100 in) long and the smallest recorded in amber (Dominican) is 0.25 mm (1/100 in) long (fig 170). Mymarommatids are very primitive and specimens recorded from Baltic amber are indistinguishable from specimens of a living species (fig 172). Fairy flies are very rare in amber, but this is probably because they are small and easily overlooked. Parasitic wasps as a whole are fairly rare in Baltic amber but fairly common in Dominican amber.

The mymarids belong to the superfamily Chalcidoidea. Some of the other members of this superfamily differ from other apocritans in that they do not have the characteristic constriction between thorax and abdomen (fig 171).

KEY TO WASPS, BEES, ANTS AND SAWFLIES

81 How many wings does it have?

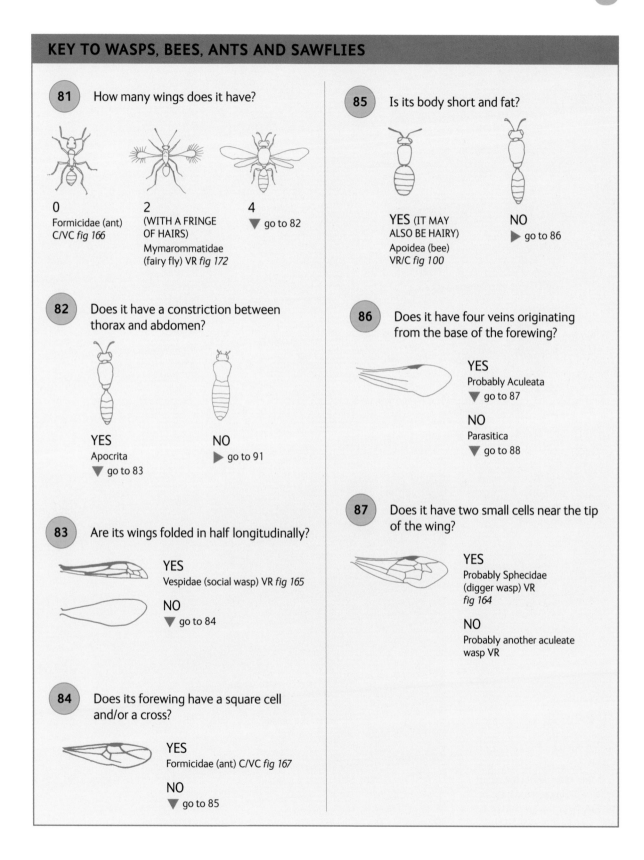

0
Formicidae (ant)
C/VC *fig 166*

2
(WITH A FRINGE
OF HAIRS)
Mymarommatidae
(fairy fly) VR *fig 172*

4 ▼ go to 82

82 Does it have a constriction between thorax and abdomen?

YES
Apocrita
▼ go to 83

NO
▶ go to 91

83 Are its wings folded in half longitudinally?

YES
Vespidae (social wasp) VR *fig 165*

NO
▼ go to 84

84 Does its forewing have a square cell and/or a cross?

YES
Formicidae (ant) C/VC *fig 167*

NO
▼ go to 85

85 Is its body short and fat?

YES (IT MAY
ALSO BE HAIRY)
Apoidea (bee)
VR/C *fig 100*

NO
▶ go to 86

86 Does it have four veins originating from the base of the forewing?

YES
Probably Aculeata
▼ go to 87

NO
Parasitica
▼ go to 88

87 Does it have two small cells near the tip of the wing?

YES
Probably Sphecidae
(digger wasp) VR
fig 164

NO
Probably another aculeate
wasp VR

KEY TO WASPS, BEES, ANTS AND SAWFLIES

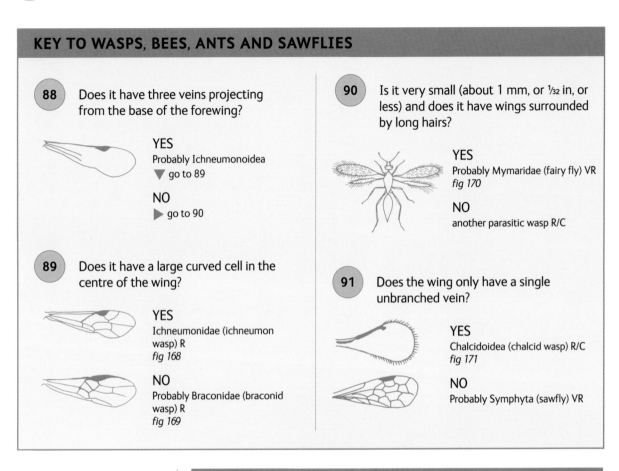

88 Does it have three veins projecting from the base of the forewing?

YES
Probably Ichneumonoidea
▼ go to 89

NO
▶ go to 90

89 Does it have a large curved cell in the centre of the wing?

YES
Ichneumonidae (ichneumon wasp) R
fig 168

NO
Probably Braconidae (braconid wasp) R
fig 169

90 Is it very small (about 1 mm, or 1/32 in, or less) and does it have wings surrounded by long hairs?

YES
Probably Mymaridae (fairy fly) VR
fig 170

NO
another parasitic wasp R/C

91 Does the wing only have a single unbranched vein?

YES
Chalcidoidea (chalcid wasp) R/C
fig 171

NO
Probably Symphyta (sawfly) VR

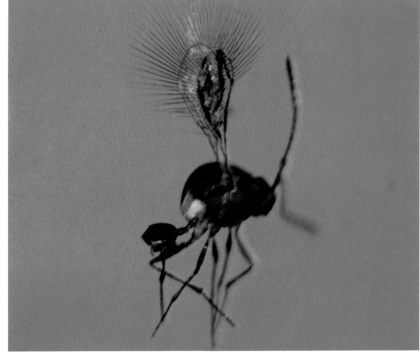

RIGHT (fig 172) Fairy fly (Parasitica: Mymarommatidae) in Baltic amber. This fairy fly is a species still living today. (Length 0.7 mm or 1/32 in.)

BEETLES (COLEOPTERA)

Insects in the order Coleoptera comprise the beetles. They are characterized by having a pair of highly toughened wing cases (elytra) that meet down the middle of the body when at rest. This is the most diverse group of insects living today with about 400,000 described species and there are also many undescribed species. Beetles as a whole are fairly common in amber and more than 70 families have been recorded; however, each family is generally rare. In Baltic amber they often have a white coating, which unfortunately hides parts of the beetle and makes identification difficult. Furthermore, most families of beetles appear very similar to each other and are difficult to identify by a non-expert. Most beetles in amber are only a couple to several millimetres in length. Larger beetles are extremely rare, presumably because they are strong and pull themselves out of sticky resin.

There are four suborders of Coleoptera, three of which occur in amber. The majority of beetles belong to the Polyphaga.

POLYPHAGA

The Polyphaga have diverse feeding habits. Given that there are so many similar-looking families recorded in amber, only four distinctive superfamilies and families are described here.

The most common beetles in Baltic amber belong to the superfamily Elateroidea, which includes the click beetles. They have long bodies with pointed and striated elytra (wing cases) (figs 173, 178). The corners of the thorax are pointed where they meet the elytra. Elateroids are wood-boring beetles that feed on rotting or diseased wood. They are very rare in Dominican amber.

BELOW (fig 173) Click beetle (Coleoptera: Elateroidea) in Baltic amber. (Length 8 mm, or 5/16 in.)

ABOVE (fig 174) Rove beetle
(Coleoptera: Staphylinidae) in Baltic
amber. (Length 2.3 mm, or 1/16 in.)

RIGHT (fig 175) Flat-footed beetle
(Coleoptera: Platypodidae) in
Dominican amber. (Length 3.2 mm,
or 1/8 in.)

LEFT (fig 176) Rove beetle
(Staphylinidae) in Dominican
amber. This one would have lived
inside termite nests. (Length 2 mm,
or ¹/₁₆ in.)

The most common beetles in Dominican amber belong to the family Platypodidae. They are known as flat-footed beetles, but their feet are not noticeably flat! They have a long cylindrical body with an elongate thorax, widely spaced legs and short antennae with an enlarged terminal segment (fig 175). Platypodids are also wood-borers, which probably bored into the *Hymenaea* trees that produced the Dominican amber. They are extremely rare in Baltic amber.

The Staphylinidae are known commonly as rove beetles or devil's coach horses. They have elongate bodies with small elytra (fig 174). They look like earwigs (Dermaptera) except they do not have pincers. They have large jaws and are voracious predators that feed on other invertebrates. There are unusual specimens present in Dominican amber that are small, have a large semicircular head, short antennae and a short tapering body (fig 90, p.46, fig 176). These beetles are specially adapted to living inside termite nests. Another family with small elytra are the Ripiphoridae (see

RIGHT (fig 177) Weevil (Coleoptera: Curculionoidea) in Dominican amber. (Length 1.7 mm, or ¹⁄₁₆ in.)

OPPOSITE (fig 178) Large beetle (Coleoptera: Elateroidea) in Burmese amber. (Length 19 mm, or ¾ in.)

fig 48, p.27); however, these have distinctive branching antennae and are extremely rare in amber. The superfamily Curculionoidea comprise the weevils. They are easy to identify in that they have a snout (fig 177). Their antennae originate on the snout and usually bend in the middle. Weevils feed on plants and are rare in amber.

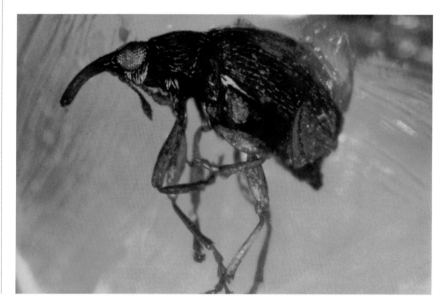

KEY TO SELECTED BEETLES

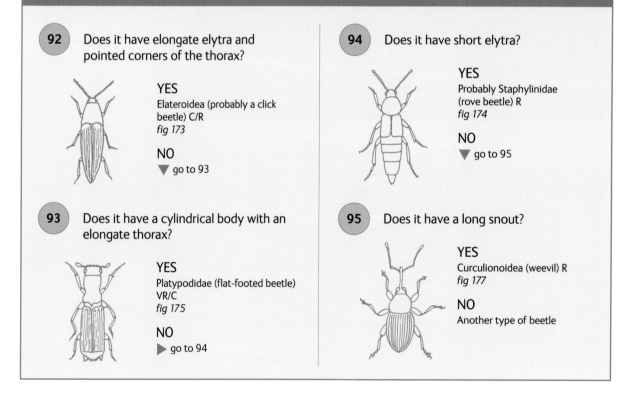

92 Does it have elongate elytra and pointed corners of the thorax?

YES
Elateroidea (probably a click beetle) C/R
fig 173

NO
▼ go to 93

93 Does it have a cylindrical body with an elongate thorax?

YES
Platypodidae (flat-footed beetle) VR/C
fig 175

NO
▶ go to 94

94 Does it have short elytra?

YES
Probably Staphylinidae (rove beetle) R
fig 174

NO
▼ go to 95

95 Does it have a long snout?

YES
Curculionoidea (weevil) R
fig 177

NO
Another type of beetle

Glossary

abdomen the hind-body of an animal, containing the organs for digestion and reproduction.

amber fossilized tree resin that is usually yellow or orange in colour and is commonly used for jewellery.

amberization the process whereby tree resin fossilizes into amber due to hardening, the loss of volatiles and polymerization.

ambroid a substance made by fusing small pieces of amber using heat and pressure. Also known as pressed or reconstituted amber.

amorphous does not have a definite form or crystalline structure.

apterygotes insects that do not have wings.

arista a bristle that projects from the antennae of some flies.

binomial nomenclature the scientific naming of organisms. Every organism is given a genus and species name in Latin.

biogeography the study of the geographical distribution of organisms.

bromeliads a family of epiphytic plants that have long stiff leaves and colourful flowers.

carapace the toughened shell of an animal, such as the head-shield of Crustacea.

caudal filaments the long, thin tails of some insects, such as mayflies.

cephalothorax the fused head and thorax of some invertebrate animals, such as spiders.

cerci a pair of sensory appendages that project from the tip of the abdomen of some insects, for example crickets.

chelicerae a pair of curved, downward-pointing fangs of some invertebrate animals, such as spiders.

chitin a tough, semi-transparent organic substance.

chrysalis see pupa.

clavus a distinctive rounded or triangular area at the base of the forewings of some insects, such as cockroaches.

colophony the solid residue from the distillation of amber to extract acid and oil.

commensalism an association between two organisms in which one benefits and the other derives no benefit or harm.

conchoidal fracture a type of breakage in which the surface resembles a shell.

copal hardened tree resin that is modern or sub-fossil. It is usually less than 2 million years old.

costa a vein of an insect's wing. It forms the front (anterior) edge of the wing.

cubitus one of the main branching veins of an insect's wing. It usually terminates on the hind (posterior) edge of the wing.

diversity the variety of life.

DNA deoxyribonucleic acid, the molecule that carries the genetic information essential for life.

ecology the study of the interactions between animals, plants and their physical surroundings.

ectognathous insects with mouthparts that project externally from the head.

endopterygotes insects that undergo complete metamorphoses during their life-cycle. The egg hatches into a larva, which pupates and metamorphoses into an adult.

entognathous insects with mouthparts that are enclosed within the head.

entomologist someone who studies insects.

entomology the study of insects.

epiphytes plants that grow attached to other plants.

evolution the gradual development of organisms from other more primitive organisms.

exopterygotes insects that undergo incomplete metamorphosis during their life-cycle. The egg hatches into a nymph, which grows into an adult.

exoskeleton the hard external structure of some invertebrates.

extant living today.

extinct died out.

fluoresce to glow when subjected to invisible light of a short wavelength, such as ultraviolet.

fossil the remains of dead organisms, usually preserved in rock.

genus a taxonomic unit consisting of a group of similar species.

halteres small knob-like organs of flies, used as balancers during flight and which are highly reduced hindwings.

hydrocarbon an organic molecule containing hydrogen and carbon.

inclusion something that is trapped in amber, such as insects, plant remains and bubbles.

infrared spectroscopy a method of analysing amber whereby infrared light is passed through a compressed pellet of powdered amber and potassium bromide. The frequencies that are absorbed are plotted on a graph.

imago the adult stage of an insect's life-cycle.

larva a juvenile crawling stage of an insect that undergoes complete metamorphosis. Also known as a caterpillar, grub or maggot.

mass spectrometry a method of analysing amber whereby a small amount of amber is vaporized. The resultant molecules are then charged and passed through a magnetic field, which changes their path according to their mass. The spectrometer then measures these changes and plots them on a graph.

median one of the main branching veins of an insect's wing, which terminates at the wing tip and/or the hind (posterior) edge of the wing.

metamorphosis the transformation of a juvenile into an adult. For insects with incomplete metamorphosis the juvenile (nymph) grows wings to become the adult. For insects with complete metamorphosis the juvenile (larva) pupates, then the tissues break down and re-form as the adult.

morphology the form (appearance) of an organism.

moulting the shedding of skin to enable an organism to grow.

mutualism an association between two organisms in which both derive benefit.

nymph a juvenile stage of an insect that undergoes incomplete metamorphosis. The nymph generally looks like the adult but is smaller and lacks fully developed wings.

organic something that contains carbon and is produced by living things.

palaeoecology the study of the relations of animals and plants to one another and their physical surroundings in the distant past.

palaeontology the study of fossils.

parasite an organism living in or on another and benefiting at the expense of the other.

pedipalps a pair of appendages attached to the head of some animals, which have either a sensory function, such as in spiders, or are used to gather food, such as the pincers of scorpions.

phoresy an association between two organisms where one is carried by the other, but is not a parasite.

plate tectonics the study of how the Earth's surface is subdivided into continental and oceanic plates that move with time.

polymer a large organic molecule composed of many smaller molecules.

polymeric something that is composed of polymers.

polymerization the process whereby molecules join together to form larger molecules (polymers).

population the inhabitants of a particular area.

proboscis the elongated mouthparts of some insects, which are formed into a tube and used for sucking up liquids, such as nectar by butterflies, blood by mosquitoes or sap by aphids.

pronotum the toughened head-shield of some insects, such as cockroaches.

pterostigma a dark spot found near the tip of the wings of some insects.

pupa a toughened case produced by insects that undergo complete metamorphosis, in which the larval tissues break down and re-form to produce an adult insect. Also known as a chrysalis or cocoon.

pyrite a mineral made up of iron and sulphur. Also known as fool's gold because of its colour.

radial sector the posterior branch of the radius vein.

radius one of the main veins of an insect's wing, which usually terminates at the wing tip. It usually runs parallel to the subcosta and branches posteriorly into the radial sector.

resin a sticky organic substance produced by trees for defence against disease and insect attack.

rostrum the elongate beak-like mouthparts of some insects, such as scorpion flies.

sclerotized toughened.

sexual dimorphism the differences in form between the males and females of a species.

solvent a substance that is able to dissolve other substances.

speciation the formation of a new species from another by evolution.

species a group of individuals with similar characteristics, which are able to reproduce with one another.

specific gravity the weight of a substance measured as a ratio of its density to either water (for solids and liquids) or air (for gases). Also known as relative density.

stellate hairs tiny star-like clusters of hairs that are commonly found in Baltic amber and probably came from the male flowers of oak trees.

subcosta a vein of an insect wing, which is usually simple and runs parallel to the front (anterior) edge of the wing.

sun spangles circular silvery cracks in amber, particularly common in modern amber jewellery. They are made artificially by heating the amber.

taxa the units of classification.

taxonomy the science of naming and classifying organisms.

tegmina toughened forewings, which are used for protection by some insects, such as cockroaches.

thorax the middle section of the body of some animals, such as insects.

type specimen the original specimen that was used as the basis for the description and naming of a species.

venation the pattern of veins in the wings of insects.

viscous a viscous substance is sticky and unable to flow freely.

volatiles liquids that evaporate easily.

Index

FURTHER READING

GENERAL

The Amber Forest
George Poinar Jr. and Roberta Poinar
Princeton University Press, 1999.

Amber, The Golden Gem of the Ages
Patty C. Rice
AuthorHouse, 4th edition, 2006.

Amber: Window to the Past
David A. Grimaldi
Harry N. Abrams/American Museum of Natural History, 1996.

Arthropods in Baltic Amber
Jens-Wilhelm Janzen
Ampyx Verlag Dr. Andreas Stark Halle (Saale), 2002.

Atlas of Plants and Animals in Baltic Amber
Wolfgand Weitschat and Wilfried Wichard
Preidrich Pfeil, 2002.

Life in Amber
George O. Poinar, Jr.
Stanford University Press, 1992.

What Bugged the Dinosaurs?
George Poinar Jr. and Roberta Poinar
Princeton University Press, 2008.

ADVANCED

Bernstein, Tränen der Götter
Michael Ganzelewski and Rainer Slotta
Deutschen Bergbau-Museum, Bochum, 1996.

Evolution of the Insects
David Grimaldi and Michael S. Engel
Cambridge University Press, 2005.

FURTHER INTEREST

Bark
Ghillean Tolmie Prance and Anne E. Prance
Royal Botanic Gardens Kew/Timber Press, 1993.

Collins Guide to the Insects of Britain and Western Europe
Michael Chinery
Collins, 3rd edition, 1993.

Complete British Insects
Michael Chinery
Collins, 2005.

PICTURE CREDITS

p.13 © AMNH; p.15 Mercer Design © The Natural History Museum;
p.19 Lisa Wilson © The Natural History Museum; p.20 Mercer Design
© The Natural History Museum; p.21 Lisa Wilson © The Natural History
Museum; p.24 © The Royal Pavilion and Museums, Brighton and Hove;
p.32 top © Rafael López del Valle, Museo de Ciencias Naturales de
Álava; p.32 bottom, p.33 top right © Malvina Lak and Paul Tafforeau,
European Synchrotron Radiation Facility; p.36 © AMNH; p.44 middle
left and bottom © George Poinar; p.49 bottom © AMNH; p.50 © John
Woodcock/Istockphoto; p.54 left © AMNH; p.55 bottom © AMNH;
pp.61-64, 70-71, 83-84, 88, 94-95, 101-102, 106, Mercer Design © The
Natural History Museum; pp.67 & 69 Mercer Design © The Natural History
Museum (redrawn from the originals by Geoff Kibby); p.85 bottom right ©
Maidstone Museum & Art Gallery.

All other images © NHMPL.

Every effort has been made to contact and accurately credit all copyright
holders. If we have been unsuccessful, we apologise and welcome
correction for future editions and reprints.

ACKNOWLEDGEMENTS

With thanks to the following for their help: Claire Mellish, Paul Taylor and
Kevin Webb.